AN OCEANOGRAPHIC MODEL
FOR THE DISPERSION OF WASTES
DISPOSED OF IN THE DEEP SEA

TECHNICAL REPORTS SERIES No. 263

AN OCEANOGRAPHIC MODEL FOR THE DISPERSION OF WASTES DISPOSED OF IN THE DEEP SEA

REPORT OF THE
IMO/FAO/UNESCO/WMO/WHO/IAEA/UN/UNEP
JOINT GROUP OF EXPERTS
ON THE SCIENTIFIC ASPECTS OF MARINE POLLUTION (GESAMP)
ORIGINALLY APPROVED AT ITS THIRTEENTH SESSION
HELD IN GENEVA, MARCH 1983

INTERNATIONAL ATOMIC ENERGY AGENCY
VIENNA, 1986

AN OCEANOGRAPHIC MODEL FOR THE DISPERSION OF WASTES
DISPOSED OF IN THE DEEP SEA
IAEA, VIENNA, 1986
STI/DOC/10/263
ISBN 92–0–125186–6

© IAEA, 1986

Printed by the IAEA in Austria
June 1986

FOREWORD

This document was originally approved by GESAMP (the IMO/FAO/UNESCO/ WMO/WHO/IAEA/UN/UNEP Joint Group of Experts on the Scientific Aspects of Marine Pollution) in March 1983 at its thirteenth session for publication as GESAMP Reports and Studies No. 19. Requests for the document from the oceanographic community have been continuous and the interest in the document has supported the original intention that its usefulness be far broader than its application to calculations of concentrations of radioactive materials in the deep sea. To ensure wider availability of the report, the IAEA has undertaken to republish the report in its Technical Reports Series with the concurrence of the other technical secretaries of GESAMP. Minor editorial changes have been made for the sake of clarity, consistency of style and to correct minor errors. The responsibility for these changes rests solely with the IAEA.

Limited quantities of potentially harmful substances have been and are being released into the marine environment, usually under the control of national or international regulatory bodies. In assessing the amount of a substance that can be released, consideration is usually given to the 'capacity' of the environment to receive the substance. The capacity itself is determined by considering the fate of the substance in the environment and the possibility that human populations or the marine ecosystem (or parts thereof) might receive amounts or concentrations of the substance at levels detrimental to their health. Such an approach to environmental protection is discussed in the GESAMP Report on The Review of the Health of the Oceans.

Recognizing the opportunity to make a substantial scientific contribution to important marine environmental problems, the eleventh meeting of GESAMP agreed to establish the Working Group chaired by G.T. Needler, on an "Oceanographic Model for the Dispersion of Wastes Disposed of in the Deep Sea" with the following terms of reference:

(1) to review the present knowledge of pathways by which substances might be transferred from a deep-ocean dumping area to man;

(2) where possible, to recommend methods for calculating the concentration of substances, arising from containers deposited on the deep-ocean floor, in the water column throughout an ocean basin; and

(3) to assess the reliability of the calculations on concentrations and, where possible, to recommend ways and means by which these might be improved.

The above terms of reference were interpreted as specifying the following four objectives for the work:

(i) Review the present knowledge of oceanic processes that may transfer substances from a deep-sea dump site back to man or his food chain;

(ii) Review methods and models currently available for estimating or calculating concentration distributions of contaminants arising from releases from deep-sea dump sites, if possible to make further developments or suggestions for developments, and give recommendations as to the currently most appropriate models;

(iii)   <u>Assess the reliability</u> of the concentration distributions
         obtained using these models;

(iv)   <u>Recommend areas for further improvements</u> and identify research
         needs.

The IAEA, as lead agency, has provided administrative and technical
support for the work of the group, which has also received support from
IMO, UNESCO, and UNEP. The Working Group met five times from 1980 to
1982 and consisted of G.T. Needler (Chairman), J.K. Cochran, J. Edmond,
G.J.R. Garrett, G. Kullenberg, N.R. Merrett, Y. Nozaki, G.T. Rowe,
J.G. Shepherd, and S.A. Thorpe.

ACKNOWLEDGEMENTS

The Working Group gratefully acknowledge the support and assistance
of many of their scientific colleagues. Thanks are due to those who
reviewed a draft version of the report and provided the Working Group
with some insight into its strengths and weaknesses. Especial thanks are
due to S. Fowler, B. Hargrave, M. Marietta, R.J. Pentreath, A.R. Robinson
and D. Wright, who attended some meetings, supplied written material,
and/or participated in the writing of the final report. Their help was
greatly appreciated.

CONTENTS

1. INTRODUCTION 1

2. GENERAL REQUIREMENTS FOR OCEAN MODELS FOR CONTAMINANT TRANSPORT . 2

    2.1. General 2
    2.2. Ocean models and parametrization 4
    2.3. Near-field and far-field models 7
    2.4. Expectation and fluctuations 7
    2.5. An analogy: The smoke-filled room 8
    2.6. Pulsed and maintained releases 9
    2.7. Conclusions 10

3. OCEANOGRAPHIC PROCESSES RELEVANT TO DEEP-SEA CONTAMINANT
    TRANSPORT 11

    3.1. Physical oceanographic processes 11
         3.1.1. Large-scale effects 11
         3.1.2. Bottom boundary layers 13
         3.1.3. Stirring, mixing, eddies, lenses, and fluctuations .. 16

    3.2. Geochemical processes 18
         3.2.1. Elements of scavenging processes 18
         3.2.2. Representation of scavenging processes 19
         3.2.3. Implications for total modelling effort 21

    3.3. Biological considerations 23
         3.3.1. General 23
         3.3.2. Mass transport processes 23
         3.3.3. Food chains and pathways 24
         3.3.4. Ecosystem damage 27

    3.4. Summary 27

4. SURVEY OF EXISTING AND POTENTIAL MODELS 27

    4.1. Near-field models 28
         4.1.1. General 28
         4.1.2. Point source 28
         4.1.3. Simple analytic finite ocean diffusive models 28
         4.1.4. Simple plume solutions 29
         4.1.5. Plume solutions in finite oceans 29
         4.1.6. Simple numerical models 31
         4.1.7. More complex numerical models 31
    4.2. Far-field models 31
         4.2.1. General 31
         4.2.2. The well-mixed box 32
         4.2.3. The Simple Finite Ocean Diffusive Model 32

4.2.4. One-dimensional models 32
4.2.5. A hybrid vertical scavenging model 32
4.2.6. Coarse box models 33
4.2.7. Plume solutions in finite oceans 33
4.2.8. Numerical models of moderate complexity 33
4.2.9. Complex numerical models 34

4.3. Conclusions 35

5. RECOMMENDED MODELS 36

5.1. Processes to be modelled 36
5.2. Definition of domains 37
5.3. General results affecting model selection 38
5.4. Model selection 39
5.5. Model sensitivity and reliability 41

6. FUTURE RESEARCH NEEDS 42

6.1. Processes needing research 42
6.1.1. The geochemical interactions in the water column..... 42
6.1.2. Vertical mixing and the thermohaline circulation
        of the oceans 42
6.1.3. Horizontal mixing and advection 43
6.1.4. Biogeochemical processes in sediments 43
6.1.5. Oceanic distributions of geochemical tracers 43
6.1.6. Quantification of biological processes 44

6.2. Models for research purposes 44

APPENDIX I: GEOCHEMICAL PROPERTIES AND OBSERVATIONS
                OF THE DEEP OCEAN 47

1. Particle distributions and fluxes 47
    1.1.   Concentrations of suspended particulate matter 47
    1.2.   The vertical flux of particles 47

2. Distributions of trace and radioactive elements 49

3. Scavenging of reactive elements 53
    3.1.   Chemical scavenging in the ocean 53
    3.2.   Determination of scavenging rate constants 60
    3.3.   Scavenging regimes 64

4. Chemical interactions in the sediment column 68

APPENDIX II: NATURAL HISTORY OF THE OCEAN 72

1. Epipelagic (= photic) zone 72
2. Mesopelagic zone 75
3. Bathypelagic zone 76
4. Benthopelagic zone 76
5. Benthic zone 78
6. Horizontal zonation and distribution of existing fisheries . 79

APPENDIX III: QUANTITATIVE ESTIMATES OF VARIOUS BIOLOGICAL
                 PROCESSES 81

    1. Calculations of maximum biological transport 81
        1.1. Transport by swimming fish 81
        1.2. Potential vertical exchange based on carbon budgets . 81
        1.3. Release of reproduction products 85

    2. Fertilization effects 86
    3. The artificial reef effect 87
    4. Calculation of the horizontal area required
        to support deep-sea fish production 88

APPENDIX IV: A SIMPLE FINITE OCEAN DIFFUSIVE MODEL 91

APPENDIX V: PARAMETRIZATION OF BOUNDARY SCAVENGING PROCESSES 94

APPENDIX VI: VERTICAL, ONE-DIMENSIONAL, ONE- AND TWO-LAYER
                BOUNDARY SCAVENGING MODELS 103

    1. A one-layer model with boundary scavenging 103
    2. A two-layer model with boundary scavenging 104
    3. A one-dimensional model with interior
        and boundary scavenging 106

APPENDIX VII: THE EFFECTS OF STRONG LOCAL SCAVENGING 109

    1. Possible modification of ocean inventory 109
    2. A simple three-dimensional model with boundary scavenging .. 111
    3. The effect of interior scavenging 112

APPENDIX VIII: A TWO-DIMENSIONAL OCEAN DISPERSION MODEL 116

APPENDIX IX: A HYBRID VERTICAL SCAVENGING MODEL 118

    1. Formulation of the model 118
    2. Application of the model to the North Pacific 126
        2.1. Natural radionuclides 127
        2.2. Stable elements 132
    3. Application to contaminants with bottom sources 137
    4. Conclusion 143

APPENDIX X: ESTIMATION OF CONCENTRATIONS IN FOOD CHAINS 144

    1. Concentration by organisms 145
    2. Multiple link food chains 146

REFERENCES 149

GLOSSARY 159

LIST OF BASIC SYMBOLS USED 163

LIST OF WORKING GROUP MEMBERS, SECRETARIAT AND MEETINGS 165

1. INTRODUCTION

This report reviews the present knowledge of oceanic processes by which substances might be transferred from a deep-sea dump site back to man or his food chain and recommends pragmatic ways to calculate such transfers in order that deep-sea dumping of contaminants may be regulated effectively.

The oceans are influenced by a variety of complex biological, geochemical, and physical processes, many of which may control the transport of a single substance and most of which are only partially understood. The calculation of transport rates to great accuracy is thus usually not possible. It is however often possible to put certain bounds on the transport rate in a particular situation so that regulations can be made for the protection of man and the environment. This report strives to identify all those processes which are known to be potentially important for the oceanic transfer of substances and to determine which of these are the most significant. In some cases these processes are well understood oceanographically, and relatively accurate ways of calculating the transfer of contaminants by them are available. In other cases, only crude estimates can be made.

Since oceanographic science is advancing rapidly, this report cannot provide an account of oceanic processes and models that will not be subject to revision in the relatively near future. The report does, however, attempt to provide prescriptions on how to calculate transfer rates in ways which are practical and adequate for many regulatory purposes and which are consistent with our oceanographic knowledge. An attempt is also made to identify the areas of greatest uncertainty (as currently seen) and to identify directions in which further effort could lead to substantial improvement in hazard or impact assessment. It is to be hoped that the report will serve to increase the understanding of the extent to which oceanographic knowledge allows accurate assessments to be made for deep-sea dumping and to identify both areas of concern and areas where predictions can be made with relative certainty.

No single adequate model exists which describes the ocean but many models exist which adequately describe parts of it. In some cases these models are relatively simple and sufficiently accurate, and several examples are provided in Section 4 and in Appendices IV, VI, VII, VIII and IX.

The selection of the optimum set of simple models usually requires clear definition of the transfer problem to be dealt with. Not only, for example, is it necessary to know the location of disposed material and the half-life, chemical and biological characteristics of the substance involved, but one must know whether its release is continuous, episodic, or of limited duration, whether the impact is likely to be on the deep-sea environment or on man via his food chain, and whether one is more concerned with small concentrations for long times over large areas or with larger concentrations in a local region. Perhaps for no situation is this point better illustrated than for the deep-sea disposal of radioactive wastes, which has stimulated much of the modelling effort described in this report. Even for given chemical reactivity and half-life, it is clear that very different models may be needed to provide estimates of the dose to a critical group and the collective dose commitment to mankind. For example, for the dose to a critical group, models that can predict maximum or limiting concentrations, including fluctuations over some time period, may be most appropriate; whereas for the collective dose commitment one is more concerned with accurate estimates of the long-term average concentrations affecting man's food

chain.   In the latter case it may not matter much whether concentrations
are high in some regions and low in others but it will be important
whether most of the substance is immobilized in the deep-sea sediments or
is available to the waters of the upper ocean.   This report attempts to
provide prescriptions for estimating concentrations and/or transfer rates
for both such situations and to indicate as clearly as possible the
assumptions for which each prescription is suitable.   It is of course
almost impossible to anticipate all the deep-sea disposal situations for
which models may be needed and those using the report should recognize
the limitations of the analyses provided.   In some cases, the development
of other models may be more appropriate than extension of those provided.
        In the next Section, the requirements of models for contaminant
transfer are discussed in more detail.   This is followed by a Section in
which relevant background information is provided on physical,
geochemical, and biological oceanic processes important for deep-sea
waste disposal considerations.   Subsequent Sections provide discussions
of available models and recommendations on how best to model particular
disposal scenarios.   Indications of model accuracies are provided as are
suggestions of areas of oceanic research and model development in order
to provide more satisfactory estimates of transfer rates in the future.
        Much of the content of the main body of the report is based either on
information that is not available in useful consolidated form in the
published literature or on the deliberations of the Working Group
itself.   Thus, the report includes a number of appendices, some of which
include detailed summaries of oceanographic information important to the
considerations of this report and not readily available elsewhere, and
others that contain calculations and models developed in the course of
the study.   The latter are certainly not completely original either in
their format or conceptual basis and in some instances contain ideas or
formulations that require further research.   They have however been used
as aids in coming to the conclusions expressed in this report.   As such
the appendices form an integral part of the report.

2.   GENERAL REQUIREMENTS FOR OCEAN MODELS FOR CONTAMINANT TRANSPORT

2.1.   General

        The terms of reference of the Working Group called for methods
for estimating the concentrations throughout the marine environment of
contaminants released on the deep-sea bed.   Such calculations clearly
need to consider the transport of contaminant by moving water, including
both advection and mixing, and the interaction with particulate material
and living organisms, so far as these are relevant.   Calculations may
need to be carried out for contaminants with widely varying properties
(including natural lifetimes and reactivity with particulates) and for
various times during various time-sequences of release.   They need to
take account of likely fluctuations that arise since the ocean is a
turbulent fluid, and may be required to provide results appropriate to a
wide range of space- and time-scales.
        In deciding how to address this substantial problem, it is helpful to
bear in mind the purpose of such models; that is, to provide information
to be used in decisions on control measures for the protection of man and
the environment from undesirable hazards due to contaminant release.   The
methodology of such protection is most highly developed for radioactive
materials, and is codified in the recommendations of the International

Commission on Radiological Protection (ICRP). Similar procedures have been promulgated for certain other contaminants but are less comprehensive. Similar considerations to those which motivate ICRP recommendations would apply, with appropriate modifications to other contaminants; an approach discussed in the GESAMP Report on the Review of the Health of the Oceans (GESAMP, 1982).

Generally speaking, one will be concerned with the exposure of man and other living things over their entire lifetimes, and, if the effects are cumulative, the total lifetime exposure will be the most important consideration. The exact sequence of high and low exposure during the lifetime is usually of secondary importance (though not necessarily negligible) as long as acute effects are avoided by not exceeding certain limiting exposures. In the case of man this means that one would be principally concerned with average exposure over several decades. Even in the case of shorter-lived organisms, the existence of very short period fluctuations in exposure is of little interest, provided that they are correctly included in the total and therefore in the average. With the exception of the lowest trophic levels (mainly bacteria and plankton in the marine environment and stages of rapid growth of other organisms), consideration of average concentrations over periods of years to decades is probably adequate. Similarly, restrictive assumptions of zero threshold and linearity of exposure and effect (usually applied to ionizing radiation) imply that one should be concerned both with effects on many people (or organisms), exposed at a low level and on a few exposed at a high level. There is therefore a need to compute collective exposure, added up over people or organisms, as well as individual exposure. It should be noted that the quantification of sublethal effects and their importance remains a matter of research. The implication is that one should consider the development of concentrations over the whole ocean, even where they may be relatively low, and should not place emphasis solely on the regions of high concentration. Furthermore, one will be interested in concentrations over long periods of time, and cannot restrict attention to the period when releases occur, especially since many contaminants remain in the ocean for a long time thereafter.

One presumes that exposure to a contaminant carries some risk and that contaminant levels should be controlled in such a way that (amongst other things) there is no unacceptable additional risk even to the most exposed groups of people, who may have rather uncommon habits. This leads to the consideration of the exposure of critical groups via critical pathways. However, controls are generally based not on extreme hypothetical considerations of what might conceivably happen to a critical group, but on the most extreme things that are reasonably likely to happen. Thus worst-case calculations, although useful, are not the only ones required. To assess the total detrimental effect of a contaminant release for all people (or things) and over all time, perhaps in order to decide whether or not it should be done at all, or in preference to some other option, the calculation must be as realistic as possible and neither the most pessimistic nor most optimistic.

All these concepts are direct translations from ICRP philosophy (see Pentreath, 1980, for an informal account) and are discussed here only as an indication of the sort of things which are likely to be of concern, and thus the types of calculations which may be required. One further useful concept is that since it is in principle possible to determine the risk associated with a given exposure to a contaminant, protection criteria in terms of risk may be converted for practical purposes to criteria in terms of exposure – ingestion of a certain quantity of contaminant, perhaps. Although it may be the average (or total) ingestion over some long period which is of primary concern, the criteria are often for convenience phrased in terms of shorter periods of time (a year is common).

One problem that the group has found to be difficult to treat in the context of deep-sea disposal of contaminants, where much is unknown and difficult to discover, is that it is not clear what level of probability is to be taken as 'reasonably likely' in making maximizing assumptions. In fact, by making more and more extreme assumptions one can often estimate higher and higher risks, but at lower and lower levels of probability, and it is not at all clear where to draw the line. Indeed, even if the line could be drawn, extreme assumptions almost by their nature defy any realistic assessment of probability. This seems to be an unsolved problem in the development of protection criteria, of particular importance in poorly understood systems such as the ocean.

2.2.   Ocean models and parametrization

The principal tools for the calculation of concentrations of contaminant in the marine environment as required by the group's terms of reference are mathematical models, of various degrees of elaboration.

The selection of appropriate models is something of an art, and usually involves trial, error, and judgement. It is, however, a general principle that a model should be no more complex than is necessary. Where simple but approximate models indicate potentially hazardous situations it may however be necessary to use more complicated but more realistic models in order that the best possible methods may be employed to assess the hazards involved and permit satisfactory regulation.

There are a number of reasons for avoiding model complexity. For example, a complex model usually requires the specification of more parameters and, if these are not known, the model's results, although detailed in nature, may be inaccurate or incorrect. It is also difficult to present and comprehend the results of complex models since, among other things, it is often hard to tell which processes included in the model or which assumptions used in its construction are actually important and which are not. Thus, a user may have trouble understanding the range of validity of a complex model and a spurious impression of reliability can easily be produced. (This can however also arise with simplistic models.) Complex models may be more difficult to interface, and time-consuming to use, when it is necessary to combine their results with other, possibly equally laborious calculations. In general one learns much more about a problem from the successes and failures of simple models, and their successive elaborations, than by the direct construction of a complex one. Simple models can be used to provide simple assessments of the importance of new effects, with the behaviour of the results usually being clear enough for the model's sensitivity to internal parameters to be rather obvious. They can often determine whether a particular process has any quantitative importance and is thus worth while including in some more complicated way. Similarly, complicated models should when possible be related to simple models (often, and preferably, analytical), both as a check on the output of the former and as a test of the importance of the extra features of the complicated model.

There are many existing models of the various components of what we have called the marine environment. Models describing aspects of the physical oceanography of the interior ocean are the most common. Various

4

models for the particulate phase and biota also exist, but seldom include more than rudimentary physical processes even though these may be important for the system being considered.

No single ocean model, whether analytical or numerical, will ever be able to represent the full range of oceanic motions. These have spatial scales from 1 mm to 10^4 km (that is, a range of 10^{10}) and an associated range of time-scales. Any computer model must by necessity have a grid scale larger than the scale of many oceanic phenomena (such as those responsible for diapycnal mixing) and so must parametrize the effect of these in terms of external variables or properties of the larger-scale resolvable part of the flow.

Parametrizations of sub-grid-scale processes are usually based on direct measurements, and/or studies of their dynamics and often are refined by matching model output to observations of large-scale features. Direct measurements (for example of diapycnal mixing) are often incomplete in space and time, and hard to relate to external, or large-scale, variables, and process studies (e.g. of internal waves, double diffusion or boundary mixing) are incomplete. Thus, modellers often have no choice but to refine a model's sub-grid-scale parametrization by comparing its output to available large-scale observations, a process that tends to be non-unique.

This last comment deserves further emphasis. A model that has been 'tuned' (that is, had its parametrization of unresolvable processes adjusted so that it describes very adequately the existing situation as well as it is known) often has very little predictive capability for other space- and time-scales and regions. For example, a model that reproduces the early stages of the spread of a tracer with a surface input (such as tritium from nuclear weapons testing) may be of very low reliability in predicting either the longer-term penetration of the tracer into the deep sea or the fate of a tracer released at the sea floor. Thus the choice of large-scale quantities against which a model is checked, and the choice of parameters for representing sub-grid-scale processes, needs to be done with great care. One technique for modelling sub-grid processes to which we shall resort is to use 'eddy diffusion coefficients', that is, to model a turbulent flux by the product of a coefficient (usually denoted by K, perhaps with a subscript to indicate direction, Horizontal or Vertical) and a concentration gradient. This is only an approximate representation of turbulent transport processes, which are inherently non-local, so the indiscriminate use of eddy diffusion coefficients is undesirable. They do however permit the computation of suitably defined ensemble-averaged or spatially and temporally averaged quantities. They are of great value in obtaining solutions to otherwise intractable problems, provide insight and qualitative results, and quantitative results under appropriate conditions. We would stress that great care is needed in the selection of coefficients appropriate to a particular process or scale. The question of geochemical "analogues", useful in tuning the coefficients, is addressed elsewhere in this report.

An objective approach to the problem of bounding the range of possible mean flows and mixing rates in the ocean, using observed distributions of tracer concentration, is being pursued by Wunsch (1981) and Wunsch and Minster (1982), and others, and, in general, yields distressingly large ranges. Undoubtedly these ranges can be reduced as better understanding of small-scale processes is obtained or as the methodology improves, especially so that all relevant data can be adduced. It is clear however that at present many of the parameters that are necessary for ocean models are not well known.

Even if the present state of the world's oceans could be well described by ocean models, it is quite possible that the parametrization and validity of these models would be inappropriate in the not very distant future (a few hundred or a thousand years) as changes in global climate occur. For long-term projections it is therefore usually preferable to use rather coarse, and hopefully robust, models that are not too sensitive to change of details of the parametrizations involved.

All these comments make it clear that it will be necessary to check the output of any model for sensitivity to the processes and parametrization built into the model, and that model output (of contaminant concentration, for example) should be considered as giving a range of possibilities.

If model output is sensitive to the parametrization of a poorly known process, it may well be that an elaborate ocean model does not produce a significant improvement in accuracy of predictions over a very simple (possibly analytic) model. Very often this situation can be predicted in advance and it is clear that the use of a complicated model can be delayed until the basic parametrizations are better established.

Between the extreme types of model (say a uniformly mixed box at one end of the spectrum, and a global, multilevel eddy-resolving model at the other) there is a range of possible models. While research on all types is probably advisable and will go on for other reasons, the optimum practical models for use in the 1980's are probably the intermediate type of model in which a site-specific sea-floor release can permeate through a variety of advective and diffusive mechanisms into the ocean interior. This approach has been used by Kupferman and Moore (1981) in a principally qualitative way.

Although it is desirable to use the simplest models that will serve the purpose, thorough investigation of the terms of reference will sooner or later require models of the oceans of the level of complexity that allow some reasonable spatial resolution, mixing and movement of water, and incorporation of the processes of scavenging by particulates and biota, simultaneously on space- and time-scales that are known to be important. Such models would, if all the relevant parameters were known, enable the calculation of the concentration field of a contaminant in the marine environment, and could be used as the substrate on which the models of marine food chains (etc.) needed to calculate the exposure of man to the contaminant in question could be overlayed. It is very likely that several such models will be needed, in order to cope with the very wide range of contaminant properties and dumping situations that will have to be considered.

One initial attempt at a model of this versatility is represented by the highly idealized six-box elementary model constructed by the NEA Seabed Working Group (1982), although this is essentially an educational caricature rather than an operationally useful model. Some of the models presented in this report may be thought of as representing steps in the same direction. The development of more elaborate models of this type is desirable, even if just to check the validity of simplifying assumptions which have to be made in their absence. The principal problem is the selection of one or more basic model types which enable the processes of interest to be adequately represented, without unnecessary complication.

In order to guide the choice of an appropriate set of models, a fairly comprehensive selection of those models (or types of models) known to us is reviewed in Section 4. From the advantages and disadvantages of these it is possible to arrive at a preliminary judgement of the types of models on which it would be worth while to invest further effort. Before doing so, and considering in detail what is known of relevant

6

oceanographic processes (Section 3), it is useful to review certain general aspects of the modelling process.

2.3.   Near-field and far-field models

In general one expects to find a region of relatively high contaminant concentration in the immediate vicinity of a source. The description of this singularity (mathematically speaking) in the near field demands the resolution of a rather wide range of small space-scales - maybe from metres to (say) a hundred kilometres. In this near-field region the concentration is likely to depend only weakly on the gross features of the rest of the ocean (size, residence time, etc.). Conversely the expected (see Section 2.4) concentration field far from the source is usually pretty much independent of the details of the near-field distribution. Thus, there is a fairly clear and rather useful cleavage of the problem into the calculation of near-field and far-field concentrations, since the details of each are often independent of all but the gross features of the other.

This is fortunate because generally speaking the mathematical methods appropriate for the one problem are poor for the other (although there are exceptions). The discussions that follow (for example, Sections 4 and 5) are therefore for convenience organized around separate consideration of models for the near and far fields.

2.4.   Expectation and fluctuations

The ocean is turbulent. The concentration field of a contaminant resulting from a release will therefore be unsteady. It may be characterized at any point by the expected value (as a function of time, release rate history, etc.) and the likely extent of fluctuations about the expected value. Fluctuations due to turbulence are likely to be large in regions with large concentration gradients, and therefore to be particularly significant in the near field. For other sources of variability, for example changes of general circulation correlated with climatic changes, this is not so.

The extent to which the problem of estimating fluctuations can be addressed depends on the type of model used. Some, for example eddy-resolving numerical models, automatically produce fluctuations by their very nature. Conversely, diffusion-advection models inherently eliminate some fluctuations because, as discussed in Section 2.2, they parametrize turbulent mixing as (steady) diffusive processes, although perturbation methods can be used to produce unsteady flow fields and therefore fluctuations. Neither type of model will produce the large-scale, long-term, climatic-change type of fluctuation unless specifically forced to do so through changes of external forcing, etc. Nor will they automatically produce estimates relevant to the quantification of the rare but extreme events discussed earlier. Generally speaking, special limiting case calculations are required to address this point.

The assessment of fluctuations in concentration fields is in its infancy. It is clear that the answers to the question 'how big a concentration might occur' depend inter alia on: the mechanisms responsible, and the space-averaging and the time-averaging scales of interest. The methods of assessment will also depend on the extent of interest in the extremes, especially since more and more extreme events or concentrations are generally expected to occur at progressively lower and lower probabilities.

In addition to these technical points, we have pointed out in Section 2.1 that a consensus has not been reached on how to take low-probability extreme events into account when taking decisions on the acceptability of disposal options for radioactive waste. The proper characteristics of models for extreme events may not be clear until such a consensus is reached.

2.5. An analogy: The smoke-filled room

The ideas discussed above concerning the concepts of the concentration field and its development with time, the near field and the far field, the effects of scavenging and of fluctuations, may all be illustrated with the help of a simple (and rather close) analogy, that of the smoke-filled room.

Consider a closed room occupied by several people, of whom one lights a cigarette. A plume of smoke is carried away by currents of air, and the plume will twist and turn so that the concentration at any point in the vicinity of the smoker fluctuates considerably. At first only those people close to the smoker (that is, in the near field) are affected by his antisocial act.

However, the turbulent currents of air gradually mix and carry the smoke to the farthest extremities of the room (the far field), and a general background level of smoke begins to build up. However, remnants of the concentrated plumes of smoke exist everywhere, although their concentration has been reduced by mixing. After a while the concentration of smoke will cease to increase so rapidly, as the processes removing it from the air in the room, such as removal by ventilation and deposition on surfaces (such as fabrics and the insides of other people's lungs) begin to take effect. The actual quantities of smoke removed are likely to increase as the concentration increases – they are approximately first-order removal processes.

Ultimately, if the smoker persists for long enough, a more or less steady state develops. This has a background concentration field determined largely by the location and strength of the source; and the location and strength of the removal mechanisms is distorted somewhat by whatever persistent currents of air may exist in the room. The concentration of smoke is always highest in the vicinity of the smoker: the near field. Superimposed on this are fluctuations of various sorts: some due to the fluctuating position of the plume of smoke at any time, some due to unpredictable fluctuations of air currents bringing concentrations from elsewhere. The magnitude of the fluctuations (expressed maybe by the peak to mean concentration ratio) is decreased if one averages concentrations over any appreciable region of space or time. Indeed, if one's measurement of concentration consists of averages over large enough regions of space or time, the fluctuations become imperceptible. Finally, the people near to the smoker notice the effects sooner than those further away, and find that the concentration in their vicinity becomes approximately steady sooner.

All these aspects mimic many features of the expected behaviour of a contaminant in the ocean. Although the ocean is more complicated due to the effects of stratification, rotation, chemistry, biology, etc., in a descriptive sense the processes of dispersion, and the fluctuations and the steady state in the near and far fields, remain the same.

2.6. Pulsed and maintained releases

One expects wastes to be dumped in the deep sea in such a way as to give a variable release of a contaminant with time. Thus, one is likely to require models that are capable of producing results on concentration as a function of time for an arbitrary time-sequence of contaminant release, that is, time-stepping models. This means in practice that at some stage analytical models are likely to prove inadequate and numerical models will be required.

There is nevertheless much to be learned from the study of special cases of which two are of particular interest, namely the single pulse release, and the point-source release which is constant for a long period of time.

As long as the released contaminant does not affect the transport properties of the ocean, the pulse release problem contains the essence of all problems, since any release scenario may be considered as a sequence of suitable closely spaced releases, whose effects may simply be added together. Mathematically the solution of the pulse release problem is the Green's Function for the system (this statement glosses over some problems to do with ensemble averages and the representation of fluctuations, and assumes that the system is mathematically linear, which is likely to be true unless high-order chemical reactions are important).

Except in cases where a pulse release is a good approximation to reality (for example, a single disposal with very rapid release of the contaminants from the waste form), excessive focus of effort on the pulse release is likely to be seriously misleading. Most practical situations involve releases maintained for substantial periods of time, either because waste disposal continues for some time, or because the contaminants are only released slowly from their containment, or both. At the opposite extreme one may consider a release maintained at a constant rate (see the analogy described in Section 2.5). Concentrations will build up and the expected value (see Section 2.4) (or equivalently the average over time-scales long in comparison with the longest period fluctuations or eddies in the flow) will eventually become steady, provided that the forcing of the ocean and the removal processes are themselves steady. This will occur after a period of time in which the rate of supply of the contaminant to the ocean is balanced by removal (e.g. by scavenging onto particles or sediment or by radioactive decay of contaminant).

Although in most real situations release rates will not remain quite steady, the concept of the steady-state concentration field is useful for various reasons:

(1)   Leaving aside fluctuations, it gives the highest concentrations reached in the history of a constant release, since after the commencement of a release the concentration at any point builds up steadily to its final state. Thus the steady-state situation can be used as a limiting case in considerations of the effects of a contaminant which has not reached steady state.

(2)   It is the case which most emphasizes the long-term effects of waste disposal, and forces one to consider these in detail. This is valuable since long-term effects are often of special concern.

(3)   It is often possible to deduce quite general statements valid
        for the steady state which assist one's understanding, as it
        is a special case for which simple solutions can often be
        constructed.

(4)   It often represents either a good approximation to a real
        projected release history, or a sensible worst-case
        assumption that is of considerable interest. It is also the
        situation that must be considered if future generations are
        to be ensured the same access to the oceanic capacity to
        receive contaminants.

For these reasons, if one were limited to considering only one
release scenario, it is probably the maintained release that would be
most useful and instructive to choose. An extended case, that where a
release is maintained for a substantial period of time, and then
terminated, is also of particular interest and value.

It is to be expected that the steady-state concentration field will
be approached much more rapidly in the near-field region (which
necessarily has a relatively short residence time) than in the
far field. This general observation, that may be confirmed by detailed
calculations and real observations, indicates that the steady-state
solutions are especially useful in the near field even though it has
often been traditional to think primarily about pulse releases in this
context.

As noted in Sections 2.4 and 3.2.1, climatic variations may occur
over periods less than the residence times of some materials that may be
disposed of by dumping. For these a steady-state model may not be
applicable (unless, for example, scavenging can remove the contaminant
over a shorter period – see discussion in the Appendices) and special
consideration may have to be given to modelling their dispersion.

If fluctuations of concentration within the ocean are found to be
important, models describing their nature will of course be necessary
even if the ocean and source can be regarded as steady.

2.7.   Conclusions

The general considerations discussed above are helpful when one
has to select, from all that is known of oceanography, those aspects that
are relevant to the present problem.

However, it is also important that further ocean modelling and
measurement be guided by the results of calculations carried out as a
result of the work for this report. Users of models will need to be
aware as to whether the output from a certain type of model is
appropriate to determining the concentration in the near field, the
far field, on the sediments, or at a particular location thought to be
critical. Only then will they be able to assess what types of new
observational or modelling effort will be effective in improving their
predictions of the concentration field in key environmental situations.

Finally, it must always be borne in mind that the basis for
oceanographic models often rests with rather recent discoveries.
Oceanographers are continually being surprised by new discoveries, for
example, of the deep-sea benthic communities at oceanic spreading centres
or the isolated lenses of water far from their possible origin (McDowell
and Rossby, 1978). The possibility of further surprises should be borne
in mind.

10

3. OCEANOGRAPHIC PROCESSES RELEVANT TO DEEP-SEA CONTAMINANT TRANSPORT

3.1. Physical oceanographic processes

There are many physical properties of the ocean which are important for the analysis of deep-sea waste disposal. In the next Section some of the large-scale characteristics of the interior ocean are discussed. This is followed by a Section dealing with the boundary layers that influence the injection into the ocean interior of substances released near the bottom as well as their potential transport up the continental slopes. The subsequent Section treats the processes that control the fluctuations in the concentration field of a substance about an average value determined in some appropriate manner.

3.1.1. Large-scale effects

It is important to note that most of what we know about large-scale average oceanic properties comes from observations of the distribution of materials introduced at the surface or from the effects of surface forcing by the winds or by changes in density. Comparatively little is known about the fate of substances released at the bottom or of the dynamical effects of bottom stresses or mixing on the large-scale circulation. Much has been learned, however, about basic oceanic processes in the past decade or two and much more information can be expected in the future.

Perhaps the most obvious property of the ocean is that it is stratified and that significant changes in density exist both vertically, from top to bottom, and horizontally, especially north to south. This arises from the large differences in the fluxes of heat and water vapour between the polar or sub-polar oceans and the equator that leave the waters of the surface equatorial ocean relatively warmer, saltier, and lighter than those at higher latitudes. In certain polar regions, such as the Norwegian and Weddell seas, heat losses lead to the formation of dense waters which eventually flow into the deep ocean forming the deepest densest water masses. In other high-latitude regions, such as the Labrador Sea, wintertime deep overturning of water occurs from time to time, creating water masses which overlie those described above. Finally, over much of the ocean surface, wintertime conditions lead to a deepening of the surface mixed layer and to the formation of water masses that greatly influence the properties of the interior ocean, especially on density surfaces that intersect the mixed layer. The general flow of dense water from the near-surface ocean at high latitudes into the deep ocean and its associated return flow is often referred to as the thermohaline circulation. Although it is known to cause the formation of features such as deep boundary currents carrying water masses away from polar regions where they are formed, little is known about the nature of the return flow from the depths to the surface high-latitude regions other than that it must lead to an ocean-average upwelling rate of a few metres per year.

The ocean is also driven by the large-scale atmospheric wind systems that result in what is known as the wind-driven circulation. Its most obvious features are the extensive anticyclonic gyres that exist between the equator and mid-latitudes and are driven by the mid-latitude Westerlies and the easterly Trade Winds. These gyres contain the strong western boundary currents, of which the best known are the Kuroshio in the Pacific and the Gulf Stream in the Atlantic. To the poleward side of the anticyclonic gyres usually exist weaker gyres turning in the opposite, cyclonic direction. At the equator strong zonal current systems exist; around Antarctica is the major Antarctic Circumpolar Current; and at the ocean's edges are other wind-driven current systems often associated with upwelling from various depths. Although all these current systems are wind-driven, their nature depends strongly on the degree of stratification present. Thus, the wind-driven and thermohaline oceanic circulations are coupled and cannot consistently be considered separately.

While the oceanic features described above are well known, details of the balances which maintain them are less well understood. The oceanic heat balance, for example, remains somewhat undetermined, partially because the strength of the interior ocean heat transport is not well known. Of the properties of the interior transports that are well known, perhaps the most important is that for the most part the advective and diffusive fluxes of all properties along density (isopycnal) surfaces greatly exceed those across such surfaces. This arises primarily because the oceanic stratification tends to restrict flows across isopycnal surfaces. Thus, on density surfaces the gradients of properties tend to be small, especially within distinct oceanographic regimes such as the anticyclonic gyres. This effect tends also to make the vertical profiles of many trace metals, nutrients, etc., similar over large oceanic areas even though their sources at the ocean's boundaries and the particulate matter that transports them vertically have fairly large horizontal variations.

Of importance to the deep-sea disposal of waste is the expectation that once material is released from the benthic boundary layer it will be transported more or less horizontally along density surfaces by mixing across density surfaces, spreading in a two-dimensional way, and that it will not in general approach the sea surface until it has been much diluted unless the density surface on which it lies approaches the surface. This possibility is discussed later. In spite of the fact that processes of horizontal dispersion dominate, vertical processes must inevitably influence the distribution of released material just as they influence naturally occurring properties of the ocean. Thus, on perhaps a basin-wide scale, one element of the heat balance is the vertical diffusion of heat downward and its transport upward by upwelling deep water masses. As mentioned earlier the location of this upward return flow is not well known and the existence of strong horizontal mixing makes it unlikely to be observed from the distribution of properties, even if it happens to occur in relatively local areas. Large but probably transient rates of upwelling in conjunction with strong downwelling events have been observed using floats (Voorhis and Webb, 1970). For materials released on the sea floor, a localized vertical return flow could be important if, for example, geochemical or biological processes are able to retain upwelled material along the continental slopes where localized deep upwelling seems most likely to occur. Such effects are most likely to be important for substances with radioactive or chemical half-lives up to a few decades that would decay before reaching the surface if only ocean-average vertical exchange processes were actually important. The inability of oceanographers to define the regions of vertical return flow remains one of the key problems for deep-sea disposal assessments.

Estimates of average vertical exchange processes can be obtained with reasonable accuracy from the oceanic distributions of various natural and anthropogenic properties. One can observe, for example, that nutrients and oxygen change in mid-depth waters in the Pacific as they flow

12

southward, interacting with abyssal and upper ocean waters. Numerical
models can be used to fit such distributions using various
parametrizations for horizontal and vertical mixing processes and
identify average values of these parameters to perhaps an order of
magnitude.

Since exchanges along isopycnal surfaces are of such importance, it
is important to examine the topography of these surfaces. This is
especially so if they reach close enough to the ocean's surface to come
in contact with significant biological activity or with regions of
intense vertical exchange, as may occur locally due to upwelling. The
density along a section in the western Atlantic and Central Pacific is
shown in Figure 1. One may note immediately that the deepest surfaces
that outcrop do so in the regions of the North Atlantic and Antarctica
where deep water masses are formed. Thus, for example, water of the same
density which at mid-latitudes exists at 2600 m in the Pacific and at
2000 m in the North Atlantic lies at the surface of the Norwegian and
Greenland seas. In general, the deepest oceanic density surfaces only
outcrop intermittently during deep water formation. They do however for
the most part normally reach high enough into the water column at high
latitudes to merit consideration of the effects of transport of material
along them.

Although time-dependent variations of the average oceanic conditions
discussed above will be treated later, one point is worth consideration
here. That is that our knowledge of the "average" ocean for the most
part is based on the assumption that it has not changed substantially
over times at least as long as the residence times of its major water
masses, perhaps several thousand years. That is to say, changes of
climatic conditions are ignored. Given the environmental half-life of
materials which may be disposed of by dumping, one must ask whether
natural climatic changes, or those caused by man (for example, by the
emission of CO_2) might not lead to large changes in oceanic
conditions. This question, which is only dealt with superficially in
this report, is most difficult to treat, although evidence exists that
the ocean circulation has undergone substantial changes in the past.

3.1.2. Bottom boundary layers

Between the interior oceanic circulation and the bottom, where
contaminants are released, there is generally a layer in which the motion
is turbulent and the temperature, salinity, and suspended particulate
matter are almost uniform. This forms at least part of the 'near-field'
region and as such requires special attention. The height of this
boundary layer ranges from metres to tens of metres. Inside this
boundary layer exist recognizable sub-layers in which the velocity is
reduced from its value outside the boundary layer to zero on the seabed
itself. The boundary layer seems to be more complex than a classical
turbulent Ekman Layer, possibly because of periodic forcing by
fluctuations in the overlying water. The layer thickness and turbulence
within it extend beyond the turbulent Ekman Layer thickness $0.4\ U_*/f$,
where U_* is the friction velocity and f is the Coriolis parameter. The
time needed to mix a substance released within the layer may be
tentatively estimated as

$$t = H_B^2/4K_V,$$

where H_B is the boundary layer thickness, and K_V is about

$0.01\ U_*^2 f^{-1} \simeq 10^{-5} U^2 f^{-1}$, wherein U is the current speed outside the

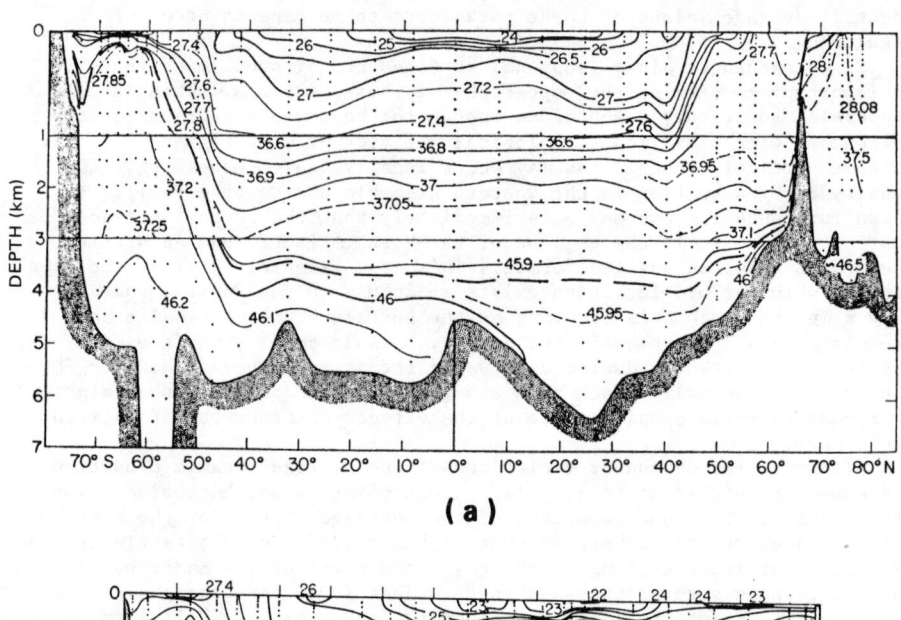

(a)

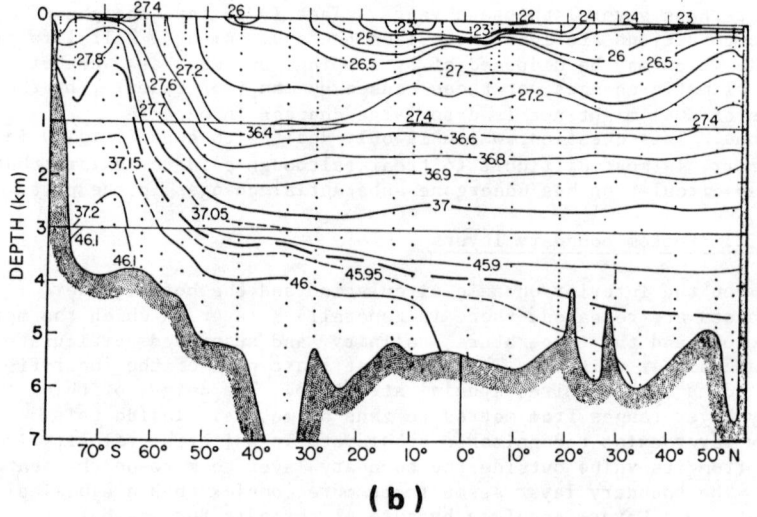

(b)

FIG. 1.   Potential density, (a) in a section in the western Atlantic and (b) in the western and central Pacific, referred to the surface (0–1000 m), to 2000 m (1000–3000 m), and 4000 m (>3000 m).   Note that the scale is expanded by a factor of 2 above 1000 m.   If flow is along true isopycnal surfaces it will almost follow the contours of potential density as shown, even though the numerical value changes from one depth interval to another (taken from Reid and Lynn, 1971).

14

boundary layer. This formulation for the eddy diffusivity K_V has not been adequately tested and should be regarded as an order-of-magnitude estimate. Typically t is about 1 - 10 days. Above the well-mixed layer the water column is normally stably stratified and vertical exchange between the bottom boundary layer and the overlying interior water column is presumably no greater than that in the interior.

Considerable horizontal mixing also occurs in the boundary layer particularly due to shear dispersion and may be enhanced by the presence of oscillating inertial or tidal currents and mesoscale events. Along the continental rise and slope relatively strong currents will generate considerable horizontal transport, probably more efficiently than over the abyssal plains.

Since the characteristics of the bottom boundary layer are by and large determined by the interior flow it is pertinent to consider separately the boundary layers over the abyssal plains and along the continental rise. Over the abyssal plains the low frequency flow is low, the friction against the bottom is weak and the bottom boundary layer is relatively thin. Along the generally irregular topography of the continental rise and slope the flow velocities are usually greater and generate stronger vertical mixing and thicker, perhaps more variable, bottom boundary layers. The mixing time in the layers will be of the same order of magnitude, that is, days, but the stronger the flow the more particulate matter that can be maintained in suspension and the greater the possibility of bottom erosion and the generation of nepheloid layers. The properties of the flow, the density distribution and the topography will also influence the exchange between the bottom boundary layer and the water above it (the interior of the ocean). Exchanges between the boundary layer and the interior ocean may be forced by time-varying exterior flow sometimes in conjunction with changing topography. Observations show that pulses of high velocity flow or fronts at the bottom may cause such changes. Recent studies in the rate of exchange between the bottom boundary layer and the interior suggest that the time-scales involved are usually of the order of months, perhaps even years, but that in some active areas they are of the order of days. Information on this matter is however limited.

The site-specific properties of the bottom boundary layer will clearly be of importance in determining the concentration distribution of a contaminant close to a site i.e. the extreme near-field concentration distribution, and its fluctuations. However, in the far field the bottom boundary layer properties local to the site will not have any significant influence on the expectation level of concentration of the contaminant. Boundary layer processes may nevertheless be important in providing cross-isopycnal mixing and thus affect diffusion even in the far field.

The concentration of suspended matter in the water column usually increases near the bottom. These so-called bottom nepheloid layers, usually contain concentration of suspended matter of about 100 μg L^{-1}, at least an order of magnitude greater than in the overlying water column. However, larger concentrations of suspended matter can occur in specific areas, at least temporarily. Recent observations on the continental rise (high-energy benthic boundary layer experiment, HEBBLE) have shown temporary concentration up to 120 μg g^{-1}. It is not clear how these large concentrations of suspended matter are generated, how persistent they are, and how frequently they occur. Nepheloid layers sometimes detach from the bottom to form midwater layers of high nephel content (see also Appendix I).

3.1.3. Stirring, mixing, eddies, lenses, fluctuations

The ocean-wide concentration of contaminants released at the sea floor can be estimated from models that allow for advection by mean currents and represent stirring and mixing processes through the use of eddy diffusion coefficients. As discussed in Section 2.4, the resulting output represents the modeller's prediction of the expected concentration field averaged over an ensemble of different realizations of the flow, or alternatively, averaged over space- and time-scales considerably larger than those associated with the stirring processes, such as mesoscale eddies in lateral directions, and breaking internal waves in the vertical. This expected concentration field is, of course, sensitive to the parametrization of sub-grid-scale processes. As emphasized in Sections 2 and 4, this parametrization is generally very uncertain. If eddy diffusion coefficients are used, appropriate values and their dependence on location and/or external parameters are needed (though they may not be of critical importance). Such values are by no means well established, though information on fluctuation kinetic energy (which may be closely related) is accumulating rapidly (Dickinson, 1983 and Richardson, 1982). Clearly then, model output in the present problem must allow for sensitivity of the predicted concentration field to poorly known parameters and may need to be presented as a range of possible values.

Quite apart from this uncertainty in the expected concentration field there will be small-scale (in space and time) fluctuations in the local, instantaneous, concentration field of a dispersing tracer. This is clear from measurements of naturally occurring tracers in the ocean (such as temperature, salinity, or other chemical properties). Vertical profiles with adequate resolutions, as well as sections and maps on isopycnal surfaces, generally show somewhat irregular patterns for the distribution of a passive property.

To some extent these fluctuations may be estimated as a "mixing length" (the length over which a quantity may be irreversibly displaced by mixing processes) times the mean gradient as discussed in Section 2.4. In the vertical such a mixing length is probably only a few metres, or tens of metres at the most. Thus, while a vertical tracer profile, or the time series of a tracer of a point, may be irregular, the instantaneous value is likely to be within the range of the mean values at positions that are at most a few tens of metres higher or lower. In the horizontal, or on isopycnal surfaces, one expects at mid-latitudes to find fluctuations associated with mixing lengths of a few hundred km, arising from the stirring by mesoscale eddies. It should be noted however that the dominant scales in space and time vary greatly from one part of the ocean to another. To some extent the mixing length approach is borne out by data (e.g. McWilliams et al., 1983), but it is becoming increasingly clear that occasional fluctuations can be substantially greater than this, associated with isolated lenses of water some thousands of kilometres from the nearest possible source, and so several years old (e.g. McDowell and Rossby, 1978; McWilliams et al., 1983). The formation, propagation, and decay mechanisms of these lenses are topics of current research in physical oceanography and still far from resolved. The lenses that have so far been found are those which originate from water (e.g. the Mediterranean water) which has a relatively clear signature. Other lenses, or their run-down remains, may be common but have not been detected because their temperature-salinity (T-S) relationship is not significantly different from the surroundings. Also unknown is the extent to which eddy fluxes of properties in the

16

ocean are carried by occasional lenses with lives of a few years and large mixing lengths, or by more transient local fluctuations with shorter mixing lengths. The possibility that these lenses may be important must be borne in mind when considering contaminant fluctuations, though it should also be recognized that they may be irrelevant if, for example, fish spend only a small part of their lives in highly contaminated lenses. In this case a biological process, fish movement, introduces a spatial and temporal averaging of the contaminant field that makes the ensemble-average model adequate. There are also grounds for believing that the lifetime of isolated lenses may be limited to a few years, due to double-diffusive and other decay processes that occur as the lens moves away from its birthplace. It thus seems unlikely that a lens would persist long enough to survive a journey from the depths of the ocean to the surface outcropping of an isopycnal and this might in any case not produce a significant perturbation of the expected concentration unless the lens originates close to the source. The results of research in this area should be kept under review.

A further complication arises in the prediction of the concentration of a contaminant with a point-like source, as opposed to substances with distributed sources such as the natural tracers in the ocean. Quite apart from the problem of lens formation, the contaminant concentration will initially be very "streaky" with local concentrations substantially greater than those predicted by a model for the ensemble average. Garrett (1981,1983) has discussed the problem within the context of two-dimensional turbulence theory, and claims that this form of streakiness will only persist for times of the order of one year and within a few hundred kilometres of the source. Other workers (Keffer and Haidvogel, 1982, Holloway, 1982) disagree. As with the problem of lenses, this topic is one of active research in which the answers are not yet well established. One should also note that this problem arising from the point-like nature of the source may be less important if the area covered by dumped material is itself large (e.g. 10^3 km^2).

The present situation may be summarized as follows:

(1)   The spatially (over a few hundred kilometres) or temporally (over a few months) averaged concentration field of a dispersing contaminant will have error bars due to uncertainties in the rates of transport processes that produce the average field, and these need to be evaluated.

(2)   There are also local or instantaneous departures from the expected field. The RMS fluctuation may be estimated in terms of a vertical mixing length of tens of metres or less, and a horizontal, or lateral, mixing length of as much as a few hundred kilometres, working on the expected gradients of concentration. However, there will be occasionally much larger fluctuations associated with isolated, long-lived, lenses of fluid that may be a few years old and thousands of kilometres from their birthplace.

(3)   These lenses are unlikely to persist for more than a few years and therefore unlikely to survive a journey to the surface, along an isopycnal, from the deep ocean.

(4)   Substantial streakiness of the tracer emitted by a point source can also be expected close to the source, but may be smoothed out within a few hundred kilometres.

(5)   Fluctuations of any sort may be insignificant if biological pathways involve large enough spatial scale or temporal averaging or if the source is widely distributed.

3.2.   Geochemical processes

The total amount of a contaminant in the marine environment
clearly depends on the strength of its sources and sinks. For deep-sea
dumping, the contaminant source is taken to be a limited region of the
sea floor. The dispersive effects of the ocean described in the last
Section tend to distribute released contaminant throughout the ocean
basin. If during this process it undergoes radioactive decay or chemical
degradation, some of it will be lost. The remainder will continue its
passage through the marine environment until it eventually reaches a sink
for the contaminant in question. For the vast majority of stable
substances the ultimate sink is burial in the deep-sea sediments.

Most substances during their lifetime in the ocean undergo a number
of biochemical and/or geochemical interactions with the suspended
particulate matter in ocean and marine sediments. Reactive substances
are thus cycled between the particulate material that primarily carries
them to the deep ocean, the deep-sea sediments where they undergo
geochemical changes and are perhaps mixed into the upper sedimentary
layer by bioturbation, and the water phase in which they are usually
carried away from the deep sea towards the surface ocean. This recycling
may occur many times before the substance is finally buried in the
sediments or is removed by radioactive decay or degradation.

The interaction of reactive contaminants with the sediments or
suspended particulate matter will modify the contaminants' water phase
concentration field from that which would be present if only physical
processes were important. In addition, the location of permanent or
temporary sinks of the contaminant and thus its availability to man or
his food chain is highly dependent on these interactions. In this
Section various important aspects of oceanic geochemical processes
relevant to deep-sea dumping are discussed. It is based to some extent
on what is known about the processes controlling the distribution of
naturally occurring elements in the marine environment. A summary of
this knowledge is given in Appendix I. In the long term the
distributions of some contaminants are to be expected to approach those
of naturally occurring species with similar chemical properties, that is,
near the steady-state situation they have natural analogues. Similarly,
models purporting to describe the large-scale long-term evolution of
reactive contaminants should be able to reproduce the basic properties of
the distribution of naturally occurring elements with similar chemistry
and reactivity.

3.2.1. Elements of scavenging processes

The process by which substances are removed from the dissolved phase
by particulate matter is generally referred to as scavenging. There are
three major steps in this process which need to be considered.

(1)   The first is the biochemical and/or geochemical interactions
        leading to association of the contaminant with the
        particulate phase, whether the particles are of organic or
        inorganic origin. If these interactions occur quickly
        enough, some sort of local equilibration of the contaminant
        between dissolved and particulate phases will occur.

(2)   The second is the transport of contaminant in the particulate
        phase. This transport will normally be very different from
        the transport in the dissolved phase, since sedimenting
        particles or faecal pellets or buoyant eggs tend to move

18

vertically across density surfaces. This disparate transport
tends to produce disequilibrium between dissolved and
particulate phases, and thus in principle the effects of
transport and sorption/desorption reactions need to be
considered simultaneously so that their opposing effects may
be balanced. The conditions under which simple approximate
models (e.g. those assuming local equilibrium) may be
adequate need to be thoroughly explored.

(3)   The third is the ultimate removal from the marine environment
(i.e. from possibility of incorporation into the marine
biosphere), by processes such as incorporation in stable
sedimentary formations, manganese accretions, insoluble
chemical forms, sedimentary burial, etc.

The first two steps, of course, modify the distribution of
contaminant within the marine environment: only in the final stage is it
actually removed. The ultimate sinks (other than radioactive decay or
chemical inactivation) are, therefore, normally at the ocean boundaries,
especially the bottom. Suspended particle concentrations are generally
large on the continental shelves and thus contaminants reaching these
regions by physical transport mechanisms will often be associated, at
least temporarily, with the coastal sediments.

3.2.2. Representation of scavenging processes

There are a number of ways that the effects of scavenging may be
included in models of contaminant transport. They are of varying
complexity and all have their strengths and weaknesses. Some are
discussed below.

Residence times. The simplest possible representation of a
scavenging process is probably just parametrization by a residence time
with respect to scavenging. Estimates of residence times may be obtained
from the study of naturally occurring analogues, and may be used in
conjunction with very simple models of the ocean (e.g. a well-mixed box)
to estimate the order of magnitude of the effects of scavenging. Such an
approach may be adequate to answer such questions as "is scavenging a
significant process, or is it trivial?", but is unlikely to give adequate
answers to more subtle questions, for several reasons.

First, an estimate of residence time usually depends on certain
implicit assumptions concerning the ocean volume and geometry, and the
substances distribution: that is, if these factors were different, the
residence time would also generally be different. The estimate is
conditional on these implicit assumptions, and a calculation based on a
residence time will only be correct if these assumptions carry over
unchanged.

Additionally, a residence time cannot express any localization of the
sinks, which may be important in altering the distribution of a
contaminant. Such a change of distribution may conceivably be just as
important as actual removal from the system: the distributions of
naturally occurring nutrients (e.g. surface depletion of silicate) by
biological scavenging processes provide an excellent example.

Lastly, the processes of scavenging, physical transport, and removal
to a sink usually have many different time-scales. Although a residence
time may adequately describe the net effect of all these processes in the
steady-state situation, its subsequent application to a transient
contaminant release is often far from justified.

Deposition velocities. Since the sinks are likely to be at the ocean margins, an alternative and somewhat more realistic parametrization of the removal process itself is provided by the use of a deposition velocity. As discussed in Appendix V, this is usually used in the form:
surface flux = deposition velocity x concentration.
Here the surface flux (quantity per unit area per unit time) expresses the rate of removal from the system to the sediments. The concentration (quantity per unit volume) is usually referred to the dissolved phase. If these are linearly related, their ratio has the dimensions of a velocity, and is a good parametrization of the scavenging at the surface.

There are several advantages of this parametrization:

(i)   the sink is localized at the ocean boundaries, which is correct for many processes of interest;
(ii)   the parametrization is closely related to the underlying sedimentary processes, and independent of irrelevant factors (such as ocean volume, depth, etc.);
(iii)   it allows easy examination of the importance of such effects as sink location and size; and
(iv)   it provides a succinct description of the boundary condition between two rather distinct elements of a complete ocean model, namely the water column and the sediments. Calculations can be made using this representation without detailed modelling of the actual processes at work in the boundary.

It does, however, depend on certain implicit assumptions, that may not be valid. These include that boundary scavenging is adequately described by a first-order irreversible reaction and that the sediments and water are in chemical equilibrium. Thus, there is no allowance in the basic formulation for remobilization of reactive substances on long time-scales and in some circumstances use of a deposition velocity would overestimate boundary removal effects. Remobilization can however be allowed for by a simple modification; namely, using concentration differences (for example, between bottom water and consolidated sediments with the sediment concentration being obtained from an additional equation describing processes in the sediments) rather than bottom water concentration as above. This is discussed in more detail in Appendix V. The main deficiencies of the deposition velocity representation are, therefore, that it can only represent that part of the modification of the concentration field due to boundary scavenging, not that due to transport by particulate matter in the interior, and that in its usual form it cannot represent remobilization of scavenged contaminant.

It is nevertheless a very simple and powerful parametrization, with which a large fraction of the effects of scavenging by many processes of interest can be represented. It has been used in the models discussed in Appendices IV, VI, VII, and VIII. These models include burial by accumulating sediments, and mixing (e.g. by bioturbation) into existing sediments (see Appendix V). The question of chemical equilibrium is examined in Appendix IX.

The estimation of values of deposition velocity from the actual processes is not, however, always simple, and unfortunately the result may depend on the actual distribution of contaminant within the sediments. When this is so an adequate model may require a more detailed representation of the sediments, allowing for the variation of contaminant concentration with depth. This might require a multilevel sediment model with the same horizontal spatial resolution as the water column model, and would lead to the two parts of an overall model being

of comparable size and complexity, which would contribute comparably to
the total labour. However, if closed form solutions for the sedimentary
component can be deduced, such elaboration may not be necessary (see
Appendix V).

Interaction between dissolved and particulate phases. Any complete
assessment of the effects of scavenging both in the interior and at the
boundaries requires that the fundamental processes controlling the
association of a contaminant with particles and dissociation from them be
represented. The actual processes involved (surface adsorption, ion
exchange, chemical coprecipitation, biological accumulation, etc.) are
many and various. Two fundamental laws of chemistry are however relevant
to the problem of developing a parametrization that is capable of
adequately representing these diverse mechanisms. One states that
reactions generally proceed faster when the reagents are present in
higher concentrations, the other that just about every reaction has an
opposing reverse reaction if you look hard enough (though it may be very
slow). These suggest that parametrization of particle-dissolved phase
exchanges as first-order reversible reactions may be a good first
approximation. The extent to which this is an adequate description
depends on the exact process under consideration. The matter is further
discussed in Appendix I.

If first-order reversible reactions are an adequate parametrization,
an important and powerful simplification is possible, because the
distribution of contaminant between the particulate and dissolved phases
will tend toward (but not necessarily reach) an equilibrium state, that
can be described by ratios of concentrations in the two phases such as
the distribution coefficient, K_d, of geochemists, or the concentration
factor of biologists. If the reactions are first-order reversible, these
ratios are independent of the actual concentration level over a very wide
range, thus permitting simplification of the modelling process.
Furthermore, if the reactions are relatively rapid, a state of
equilibrium between the dissolved and particulate phases is sometimes a
good approximation. This can make it unnecessary to model both the
particulate and dissolved phases, or compute the interactions in detail,
leading to a further modelling simplification. The first-order
reversible reaction formulation is used in the model described in
Appendix IX, where some of the modelling questions raised here are
discussed in more detail.

Finally one should note that the adoption of such an approximation
does not imply that the distribution coefficients should be universal
constants, only that they should be definable in principle: variation of
their values with physical, chemical and biological conditions would
still be allowed, and indeed expected. The values are only required to
be independent of contaminant concentration, over the range of interest.

3.2.3. Implications for total modelling effort

These considerations imply that a model of the marine environment
which is capable of fully representing just the basic scavenging
processes would have to comprise:

(a) A representation of the water column (at some appropriate
     spatial resolution).
(b) One or more parallel representations of various particulate
     phases in the ocean interior (e.g. fractions of various sizes

of both mineral and biological particles). These would
probably have the same spatial resolution as (a).

(c)   Some representation of the sediments, at the same
        (horizontal) spatial resolution as the ocean interior
        models. This may also involve the representation of more
        than one phase (e.g. interstitial water and particles) and
        several levels.

(d)   A representation of the geographic location of sinks. This
        would be closely linked to the sediment representation of (c)
        and at a minimum could include the potential for strong sinks
        on eastern boundaries (e.g. due to high productivity
        associated with upwelling, Appendix I) and inhomogeneity of
        removal in bottom sediments (linked as a first approximation
        to the particle flux to the bottom).

Considering our present limited knowledge of the details of the
geochemical processes involved and equivalent problems discussed
elsewhere in this report with modelling physical and biological oceanic
processes, it is clear that such a complete representation of scavenging
is not necessarily appropriate at the present time for operational models
of the transfer of contaminants from deep-sea dump sites. It would also
be at variance with the recommendation contained in this report regarding
the use of the simplest models consistent with the needs of the problem
to be addressed and our knowledge of the processes involved.

The inclusion of scavenging in some form in models of contaminant
transfer is however one of the most important advances that can be made
at the present time. This is especially true since it is often involved
in the details of how the contaminant enters man's food chain. Also, any
comparison of model results with the distribution of natural analogues
requires the inclusion of scavenging mechanisms for reactive elements.
Thus in the Appendices various models including scavenging in one form or
another are examined with the aim of delineating those problems where it
is important. The results of the studies have influenced the
recommendations of this report.

One aspect of scavenging that has been treated in some detail
(Appendix VII) concerns the effect of intense local scavenging of highly
reactive contaminants. The greater part of the total inventory of such
contaminants may be associated with the sediments and the concentration
in the water column substantially modified (both horizontally and
vertically) by scavenging falling particles, perhaps further enhancing
the contaminants' removal to the sediments. It is shown however in
Appendix VII that, although at some distance from the source the
concentrations are substantially modified, the sediment-water inventories
are not affected (at least at steady state and for horizontally uniform
physical and geochemical properties). Thus, intense local scavenging may
certainly lead to more of the contaminant on the sediments near the
source than would have been expected had the concentration field not also
been affected, but this does not influence the overall efficiency of
scavenging by sediments.

When scavenging is the dominant removal process, but not so strong as
to prevent the establishment of substantial far-field concentrations, the
possibility arises that the greater part of the removal may occur far
from the source, for example on shallow-water sediments. This is
discussed in Appendix I and the effect may be evaluated using a model of
sufficient complexity.

3.3. Biological considerations

3.3.1. General

The principal areas of biological production in the oceans are the
shallow waters and especially the coastal regions. The biomass in the
deep ocean is much less than in these productive regions. Nevertheless,
it is by no means a lifeless desert, but supports a substantial and
varied biota, adapted to life in reduced light, and ultimately is
dependent for its energy on organic matter from above. A brief
description of the natural history of the deep oceans is given in
Appendix II.

The presence of life in the deep oceans affects the assessment of the
transport of contaminants released there in several possible ways.
First, organisms sometimes take up contaminants, often selectively, and
create a possible hazard if ingested by man (or other animals). They may
therefore be links in food chains leading to exposure pathways. Second,
if the accumulation of contaminants, and the biomass and movements of
organisms are sufficient, the total quantities of contaminant moved
around in this way might be a significant part of the total mass
transport, sufficient to alter the concentration field which would
otherwise be established by physical and geochemical processes.
Furthermore, the movements of living things may disturb their environment
and thereby lead indirectly to mass transport of contaminants which they
do not actually carry themselves, for example the mixing of surficial
sediments by bioturbation. Containers on the seabed modify the
environment which may affect species composition and potential dispersal
of organisms, while the contaminants also may affect metabolic activity
(see Section 3.3.4).

The distinction between these two aspects of biological processes is
important, because a mechanism can only cause mass transport if the
quantities of contaminant, biomass, etc., involved are sufficiently
large. It is gross quantities which matter. On the other hand, a
significant exposure pathway may be created by a food chain which
transports only a tiny fraction of the total quantity of contaminant,
insufficient to have any discernible effect on its overall distribution.

The significant mass transport mechanisms need to be included in
whatever models are used to estimate the concentration field of a
contaminant in the marine environment, along with the physical and
geochemical transport processes, although the pathway mechanisms need
not. Their effects may be estimated subsequently, once the concentration
field has been established. As mentioned elsewhere, however, the
selection of the most appropriate model may depend on knowledge that
certain pathways are likely to be important.

Finally, as mentioned above, the presence of a contaminant may affect
organisms themselves, and possibly damage the ecosystem. Such effects
may be estimated once the concentration field is known, and may therefore
be handled in a fashion similar to that used for food chains; indeed,
food chains may be involved.

3.3.2. Mass transport processes

Biomass in the deep ocean is tiny compared with the mass of the
environment. For example, the deep-ocean standing crop is estimated to
be a few g/m^2 over depths of 1000 m of water or 10 cm of sediment, that
is, only a few grams of biomass to every thousand tonnes of water, or

several hundred kilograms of sediment. For this reason it would not be expected for mass transport of contaminants by living things to be a significant process, even allowing for possible concentration of contaminants. The question is discussed in detail in Appendix III, where a variety of calculations on possible mechanisms carried out by the Working Group are presented. These show, for example, that even under extreme conditions the horizontal transport by swimming fishes is something like a million times smaller than that of the water in which they swim. Similarly, upward transport by buoyant eggs of benthic organisms is perhaps 10^{-5} to 10^{-7} of the transport by the slow overall upward transport of the oceans, so that it cannot be significant even if there is a substantial vertical gradient in contaminant concentration (say, a factor of one hundred), and the egg production just happens to be concentrated in a tiny fraction of the ocean which coincides with the point of release of the contaminant. Further examples are discussed in Appendix III, and similar calculations have been carried out by Robinson & Mullin (1981) and Anderson (1982).

There are only two identifiable exceptions to the rule that mass transport by organisms is insignificant:

(1)   One is the mixing of surficial sediments by animals (bioturbation). This reworking continually acts to expose new particulate material at the sediment-water interface, and, in the short term (indefinitely in the case of sufficiently short-lived contaminants), greatly accelerates the scavenging (or release) of reactive contaminants at the interface. The effects of bioturbation are considered in more detail in Appendix V.

(2)   The second is incorporation of contaminants into biological (and other) materials in the surface layers of the ocean, followed by downward transport as sinking particulate material (mainly detritus) and release as the biogenic material (which may be organic, calcareous, or siliceous) dissolves or is consumed.

The latter process is of course the principal cause of the well-known inhomogeneous distributions of nutrients in the oceans, and also of many trace elements. Whilst it is of biological origin, its description falls primarily within the field of geochemistry, and it is discussed in detail in the geochemical Sections of this report (Section 3.2 and Appendices I and IX). The reason this process is relatively important is that the quantities of material involved are relatively large and ubiquitous. Over most of the ocean it is the most significant process counteracting physical processes in the other direction. One should also note that this biological flux is clearly towards the bottom in the deep ocean. Despite seasonal and diurnal migrations of organisms, no upward flux of biomass of an equivalent magnitude is known to exist.

3.3.3. Food chains and pathways

The estimation of the transfer of contaminants in food chains is an integral part of the assessment of the effects of their release, but belongs rather to the hazard assessment part of the procedure than to the estimation of concentrations in the marine environment which is the principal question addressed in this report. The general procedures required for assessing transfer by marine food chains are described in

Appendix X. The conclusion that there is no reason to suppose that concentration of a contaminant necessarily occurs at each step in a multiple food chain is of interest.

Sites for deep-ocean waste disposal will presumably continue to be located in regions not directly accessible to man. Some general features of food chains are thus also of importance, because they provide the most direct potential link between man and the near-field region of elevated concentrations, in the vicinity of the disposal site (see Section 2). Properties of food chains can determine certain spatial scales in the vicinity of the release over which near-field concentrations need to be evaluated.

A minimum region of the ocean floor is required to sustain production of organisms likely to be harvested by man. The smallest levels of production relevant to exploitation by man may range from a few hundred grams per day (0.12 t·a^{-1}) which might be consumed by an individual to levels greater than several hundred tonnes per year which might be sufficient to sustain a commercial fishery. Detailed calculations (Appendix III) indicate that the minimum areas required to sustain such levels of production are of the order of one hundred square kilometres or more. This is about the scale chosen in Section 2 to distinguish the extreme near field and near field. This implies that the detailed distribution of concentration in the extreme near field (around canisters, etc.) is of little consequence in so far as hazard to man is concerned, since the biota required to sustain a pathway must feed over a larger area, resulting in a lower concentration. The high extreme near-field concentrations are only of significance in the case of potential damage to benthic organisms.

These estimates provide a rather strong limiting case calculation: a food chain cannot be sustained by animals ranging over a smaller area. In practice the animals may range over a much larger area. How large this area may be is difficult to assess. One species of rat-tail fish (Coryphaenoides (Nematonurus) armatus) is distributed over major fractions of entire ocean basins (several thousands of kilometres). Such fish are certainly capable of swimming thousands of kilometres, and the fact that ripe females have never been found in the eastern North Atlantic suggests that they may do so. However, direct evidence is lacking. Conventional fish tagging experiments are impractical since specimens cannot be brought to the surface and returned alive. Experiments with labelled food may be feasible, but deep-sea fishing is so sparse that it will be difficult to get adequate returns. Observations of fish congregating on bait (Isaacs and Schwartzlose, 1975) show that they are quite mobile. However, even with an abundance of one fish per 1000 m^2 (Appendix III), the observations do not prove that movements of more than a few hundred metres need have occurred. Thus although deep-sea fish may range over large distances, at present this cannot be unequivocally demonstrated. Nevertheless, in a multiple food chain of scavengers and carnivores, animals at each stage probably range quite widely, eating prey that have themselves quite possibly covered substantial distances, and accumulated contaminant corresponding to some average level of concentration, rather than an isolated peak of high concentration. Thus it is unrealistic to imagine eating fish that have fed only on prey which have fed only in a region of peak concentration. Dilution with other food occurs with each transfer, and some degree of averaging is likely. Certainly experience on the continental shelves would suggest that this is so (Isaacs, 1972 and 1973).

These considerations are relevant to the hypothetical "bathys restaurant" pathway considered by the Working Group. This supposes that

TABLE 1. NEAR FIELD MODELS

Model	Resolution	Finite Ocean	Decay	Flow	Biogeochemical Scavenging	Time Development	Type	Complexity Cost	Reference
Point Source	Unlimited	No	Possible but not necessary	No	No	No	Analytical	V. Low	Appendix IV
Simple Finite Diffusive Model	Unlimited	Yes	Yes	No	No	No	Analytical	V. Low	Appendix IV
Finite Ocean Diffusive Model with Scavenging	Unlimited	Yes	Yes	No	Yes	No	Analytical	Low	Appendix VII
Plume Solutions	Unlimited	No	Yes	Steady(1D)	No	No	Analytical	Low	
Plume Solutions in Finite Ocean	Unlimited	Yes	Yes	Steady(1D)	No	Yes	Small Numerical	Medium	Shepherd (1976)
Finite Simple Difference	Low (Grid Size) (Limited)	Yes	Yes	Steady (2 or 3D)	Yes	Yes	Medium	Medium	Shepherd (1983) or Similar to Fiadeiro & Craig (1978)
Stochastic	Medium (Grid Size) Limited	Yes	Possible	Unsteady (2D)	?	Yes	Medium Numerical	High	
Eddy-resolving	High (Grid Size) Limited	Yes	Possible	Unsteady (2D)	?	Yes	Large Numerical	V. High	Keffer & Haidvogel (1982) Holloway (1982)

an entrepreneur establishes a restaurant which serves only bottom-living deep-sea fish. The cooks in such a restaurant might, if they enjoyed the food they served, have consistent high consumption rates. Such a scenario is improbable but not inconceivable. Furthermore, the source of supply of the fish might by chance coincide with the near-field high concentrations around a disposal site. Whether or not control of waste disposal should be based on such a pathway is a matter for consideration by regulatory bodies and is not within the competence or terms of reference of this group. However, it is clear that the calculation of an appropriate concentration for such a pathway would involve the near field rather than the extreme near field. Calculations for sustainable fisheries would require averages over larger areas again.

3.3.4. Ecosystem damage

Like the actual assessment of particular pathways, the estimation of damage to biota and the marine ecosystem in general has not been addressed by the Working Group. Given the methods we recommend for the calculation of concentrations from any desired scenario of contaminant release, however, the associated damage to organisms may be calculated in a straightforward manner, provided that necessary "dose/effect" data are available. Such data are lacking for deep-sea organisms, even for the relatively well-researched case of the effects of radiation (IAEA, 1976, 1979). The methods discussed in Section 5 may be used to calculate relevant concentrations as and when suitable dose effect data can be assembled.

3.4. Summary

In this Section, the physical, chemical and biological processes known to be of importance in determining the concentration of a contaminant released in the deep sea have been discussed in varying detail. It will be clear to the reader that in general, although not always, the mass transport or the concentration field of a contaminant is determined primarily by physical processes with geochemical and biological processes being of descending order of importance. This observation will be reflected in the models discussed and recommended in the remainder of this report.

The search for important mechanisms, especially biological ones, leads one to examine rather hypothetical or maximizing transfer processes. One must expect that as more is learnt about the deep sea more such possibilities will be suggested. However, it is unlikely that these will invalidate the general considerations presented in this Section, although care will have to be taken to investigate and quantify all possibilities as they arise. It is also unlikely that the complete removal of such possibilities, especially those including some complicated mixture of physical, geochemical, sediment, and biological transfer mechanisms, will ever be achieved. Bounds on their effects should however become narrower and procedures for their identification, quantification and elimination better developed by continuing research into deep-sea-oceanic processes.

4. SURVEY OF EXISTING AND POTENTIAL MODELS

In this Section the requirements of models for the transfer of substances from deep-sea dump sites are further refined and discussed.

This is done in the context of various models and calculations prepared during the study as well as on the basis of various models available and known to the Working Group. The focus will be on physical models of water movement since it is that which must be modelled as the framework on which to superimpose the effects of geochemical and biological processes, etc. The discussion is based on the idea of the sensible division of the overall problem of contaminant transport into those of the near and far field. The reader is reminded however that for certain processes the division between these regions is not always distinct and this must be taken into account when applying models to particular problems. Many methods utilized in air pollution studies (e.g. the estimate of peak near-field concentrations as a function of averaging time and the use of deposition velocities in studies of acid rain), and not described here, may have application to deep-sea studies, particularly to that of fluctuations, and should be kept under review.

4.1.   Near-field models

4.1.1. General

For the near-field high spatial resolution is required, at least close to the source. For this reason analytical models are useful, because their spatial resolution is not limited by any considerations of grid size, although earlier caveats about correct parametrization of small scales (Section 2.2) still apply. However, the inclusion of more processes of interest (for example, scavenging) makes it more and more difficult to construct analytical solutions, so there is a natural progression from very simple analytical models to more complex and ultimately numerical models. The results of the discussion below are summarized in Table I for convenience. Most of the discussion is centred on solutions for the steady state. These are easier to find and are of particular importance in the near field, where steady state (at least in a statistical sense) is achieved relatively rapidly.

4.1.2. Point source

The simplest near-field model is probably that of a point source in a static diffusive medium. If one assumes that the source lies on a reflective (non-absorbent) boundary of an infinite half-space, the solution (for a continuous source Q) is simply $C = Q/2\pi Kr$, where C is the concentration, K the diffusivity and r the distance from the source. The combination Kr may be in reduced co-ordinates (see Appendix IV) to allow for anisotropy of the medium.

This is clearly a very highly idealized model: the ocean is infinite, there is no flow or scavenging or time-history. The model is however probably adequate to give order-of-magnitude estimates for the steady state very close to the source. It shows immediately the nature of the high concentration field expected in this region.

4.1.3. Simple analytic finite ocean diffusive models

Two useful extensions of the point-source model are easily made. These are

(a)   allowance for radioactive decay (or other first-order irreversible uniform removal processes); and
(b)   provision of a boundary to create a finite ocean.

Inclusion of these aspects leads to the model given in Appendix IV.
The finiteness of the ocean allows the background concentration (a crude
representation of the far-field solution) to build up, so that one can
compare the relative importance of the near-field and far-field solutions
at various distances (normalized for anisotropy) from the source. It
shows also that the nature of the singularity at the source is not
affected by the inclusion of decay processes or the far-field solution.

The effects of an absorbing boundary, interior scavenging and finite
size of the source region, can be taken into account (see Appendix VII).

4.1.4. Simple plume solutions

If one attempts to enhance the point-source solution by including
flow, one is led to a class of models widely used in air pollution
studies (see e.g. Pasquill, 1975; Csanady, 1973) that may be called plume
solutions. In an oceanic context it is useful to parametrize the
fluctuation of advection (the 'wind') as a diffusive term, and allow for
upstream mixing (not usually necessary for atmospheric applications).
This is not difficult, at least for some representations of pulse release
dispersion. The resulting analytic solutions are sufficiently
complicated to require evaluation by a powerful calculator or small
computer, but retain the advantage of unlimited resolution.

These solutions serve to make the point that when the flow is weak
compared to mixing, the principal effect of flow is to distort the
concentration fields (i.e. move the contours) that would have been
established in the absence of flow. Only for strong flows do the contour
levels and regions occupied change dramatically. These solutions are
also useful for giving a more realistic visual impression of the
'footprint' of high concentration likely to be created in the vicinity of
a release: there is no essential difference from the familiar footprints
of pollution created by chimney plumes.

4.1.5. Plume solutions in finite oceans

If one attempts to extend the method (explicit use of Green's
Functions) used to derive simple plume solutions, so as to allow for
flow, radioactive decay, and a finite ocean simultaneously, one is led to
the type of model developed by Shepherd (1976), which was used to some
extent by the IAEA consultants (see IAEA-210, 1978). The solutions for
this model can be written down as analytical expressions, but are
sufficiently complicated (involving the summation of many terms) to
require evaluation by a small computer. It is natural to carry out the
calculation in such a way that the time-development is automatically
available, and the resolution is unlimited (in the sense that the
concentration can be calculated for any point in space, even very close
to the source, if required).

The model includes more of the essential processes of interest
(though not very realistically), and yields both near-field and far-field
solutions. In this respect it is like the model of Appendix IV, and is
indeed conceptually rather similar. Although it has previously been used
to obtain far-field results, it is in fact rather better suited to the
computation of near-field concentrations, and it is possible that this
model may find renewed application for this purpose. The only major
process left totally unrepresented by this model is scavenging. Because
of the structure of the model, it may only be possible to include this in
a crude fashion.

TABLE II. FAR-FIELD MODELS   (All have finite ocean and decay)

Model	Resolution	Flow	Biogeochemical Scavenging	Time Development	Near Field Solution	"Realistic Ocean"	Type	Coordinate System	Complexity Cost	Reference
Well-mixed box	V. Low	No	Crudely	Yes	No	No	Analytical	n/a	Lowest possible	-
Finite Ocean Diffusive Model with Scavenging	Unlimited	No	Yes	No	Yes	No	Analytical	Cylindrical Ocean	Low	Appendix VII
Basic One-dimensional	Unlimited	No	No	Feasible	No	No	Analytical	Vertical	Low	IAEA 210
Extended One-dimensional	Unlimited	No	Yes	Feasible	No	No	Analytical	Vertical	Low	Appendix VI
Hybrid Vertical Scavenging	High (1D)	Upwelling and return	Yes	Yes	No	No	Small Numerical	Vertical	Low	Appendix IX
Coarse box	Low	Steady	Yes	Yes	No	No	Small Numerical	Isopycnal if required	Low	NEA, (SWG)
Plume Solutions in Finite Ocean	Unlimited	Steady (Horizontal)	No	Yes	Yes	No	Small Numerical	Cartesian	Medium	Shepherd (1976)
Many boxes	Medium	Steady	Yes	Yes	Crude	Not very	Medium Numerical	Could be Isopycnal	Medium	Under Development
Finite Difference (2D)	Medium	Steady (Arbitrary)	Yes	Yes	Crude	Not very	Medium Numerical	Isopycnal	Medium	Shepherd (1983)
Finite Difference (3D)	Medium	Steady 3D (Idealised)	Feasible	Feasible	Crude	Fair	Medium Numerical	Cartesian (could be Isopycnal)	Medium	Similar to Fiadeiro & Craig (1978)
"Realistic 3D"	High	Steady 3D	Could do	Yes	Yes	Fair	Large Numerical	Cartesian	High	After Bryan, Semtner and Various Workers
Eddy Resolving GCM	V. High	Unsteady 3D	Could do	Yes	Yes	Not very	V. Large Numerical	Cartesian	V. High	Robinson et al. (1977); Semtner and Mintz (1977); Schmitz and Holland (1982)

4.1.6. Simple numerical models

If one wishes to construct a model capable of incorporating all the major processes of interest simultaneously, some sort of numerical model is necessary. The most obvious choice is a simple finite-difference model on a fairly coarse grid. Such models, together with the closely related box models and finite-element models, are discussed in more detail in Section 4.2 since they are rather more appropriate to the calculation of far-field concentrations. They are capable of producing useful results, however, for the fairly near field also, limited by the resolution of the grid or mesh used.

Even quite simple numerical models (for example, the two-dimensional finite-difference model described in Appendix VIII) are capable of yielding very useful and instructive results on certain questions, and have the advantages of comprehensibility and simplicity. They are, however, not capable of resolving the extreme near field and may give misleading results in special cases (for example, with extreme sediment scavenging) if used injudiciously.

4.1.7. More complex numerical models

More complex numerical models may be required if it is necessary to include, for example, real ocean topography or detailed resolution of the extreme near field. Similar considerations arise for the far field and are discussed in Section 4.2. In the near field, the main complications arise when one wishes to compute the effects of turbulent dispersion explicitly. This may be desirable if one is concerned that the advection/diffusion parametrization may give incorrect results.

A type of model required for this purpose is an eddy-resolving dispersion model, such as those of Bennett and Haidvogel (1981) and Holloway (1982). Even with efficient coding, recording of flow history for subsequent reuse, etc., these are still fairly complex to run, and are therefore probably best adapted to the provision of answers to specific questions, rather than general use. They use substantial computing time because each model run gives results only on a single 'realization', and many runs are required to assemble statistics of expected concentrations and confidence limits.

Less elaborate models neglecting the computation of true turbulence but simply imposing random forcing in the flow field have been used in estuarine research (e.g. see Allen, 1982) and might be useful in describing near-field dispersion, particularly in regions of complex topography. It is not clear however that the problems of interpretation would be reduced. These models are however noted in Table I as 'stochastic' models, for completeness.

4.2. Far-field models

4.2.1. General

Far-field models are distinguished by the fact that the finiteness of the ocean is likely to be important, but the details of the near field are not. The progression in complexity for this case is essentially from low to high in the spatial representation, with ever increasing realism. The main points are summarized in Table II.

4.2.2. The well-mixed box

The simplest model of a finite ocean is a single well-mixed box. Such a model clearly has no spatial resolution. It has no representation of flow and mixing processes, being equivalent to the assumption that these are much more rapid than any other processes of interest. This is clearly false for short residence time contaminants, but the model is nevertheless very useful. This is because it is easy to show that if the only removal processes are uniform first-order irreversible (such as radioactive decay) the well-mixed box gives the correct answer for the steady-state oceanic average concentration. Thus, for weakly reactive contaminants, or those where removal can be adequately parametrized by a residence time, this very simple model gives useful order-of-magnitude results which can be used as scaling values for the results of more elaborate calculations.

4.2.3. The Simple Finite Ocean Diffusive Model

The Simple Finite Ocean Diffusive Model (see Appendix IV) contains what is essentially the well-mixed average concentration with a near-field representation superimposed. As a far-field model it therefore has little advantage over the single box, but is mentioned here for completeness.

The models of Appendix VII, which allow for interior and boundary scavenging, are particularly useful in determining whether most of the contaminant will be removed within a region that is much smaller than an ocean basin, thus making the far-field concentration very small.

4.2.4. One-dimensional models

Much of the interpretation of geochemical tracer distributions has until recently been done using one-dimensional (vertical) models. For some tracers (e.g. temperature) such models are capable of giving quite good results, and they do permit rather detailed treatment of scavenging processes within the context of analytical solutions (hence, perhaps, their popularity). Such models are also capable of giving quite good results on the overall vertical distributions to be expected from artificial contaminants under some circumstances (see Appendix VII). However, they never give an adequate representation of the near-field distribution. Although they often include vertical velocities, usually the vertical flow has not been accounted for in a consistent way as when the model fails to include a return flow of downward-moving water carrying the substance of interest and hence contributing to the net vertical flux.

Time-dependent results from such models are likely to be unreliable for some time-scales. A model of this type, including only diffusive fluxes, was used as the basis of IAEA-210, since it was found to give a satisfactory representation of the results of Shepherd's 1976 model under the circumstances of interest. In that case, separate calculations relevant to the near-field problem were also required.

4.2.5. A hybrid vertical scavenging model

The model given in Appendix IX is an extension of the one-dimensional models discussed in Section 4.2.4. It avoids some of the problems associated with the total flux considerations in one-dimensional models and introduces some realistic aspects of sedimentary interactions. As such, it allows comparisons of model results to be made with observations of the distribution of substances already in the ocean, and, within its assumptions, allows estimates to be made of long-term horizontally averaged distributions for reactive substances released from a dump site.

4.2.6. Coarse box models

If one wishes to represent flow and mixing in a system of low spatial resolution a coarse box model may be appropriate. A six box model is being studied by the Systems Analysis Task Group of the NEA Seabed Working Group (1982) in the context of high-level radioactive waste disposal, as a very simple example of the sort of model they may ultimately wish to use.

Such a model clearly can give no satisfactory representation of the near-field distribution, but may give good results for long-lived, weakly reactive contaminants with long-maintained releases, the central problem for which it is being developed. It is easy to incorporate scavenging and biological processes, the box geometry may be constructed in an isopycnal frame of reference and, since the model is small, the numerical calculations required are minimal.

Since however the flow and ocean geometry are necessarily highly idealized, and it is difficult to specify the exchanges between the components, it is likely that such simple box models should be reserved for educational purposes rather than operational use.

4.2.7. Plume solutions in finite oceans

The 'plume in a box' model of Shepherd (1976) (see also Section 4.1.5) yields a concentration field throughout the (idealized) ocean volume. It was used as a basis for IAEA-210. It is of interest (like the Simple Finite Ocean Dispersion Model) because it yields some sort of solution to both near-field and far-field problems simultaneously. However, the geometry and flow field are so idealized (vertical motion is forbidden) that its utility as a far-field model is limited. It is also not possible to incorporate scavenging processes at all easily, and it does not make a good basis, therefore, for the sort of general-purpose calculations we envisage as necessary.

4.2.8. Numerical models of moderate complexity

If one wishes to represent simultaneously the effects of flow, mixing, decay, scavenging, and biological transport some sort of numerical model in two or three dimensions is required. This might be a fairly elaborate box model (say 30 or more boxes), a finite-difference model, or a finite-element model. There is indeed no essential difference between these various types, although there may be technical advantages in using one type or another for particular problems. The spatial resolution may be made as great as desired (at ever-increasing computing cost), and there is no difficulty in representing the features of the ocean topography at the appropriate resolution. It is possible to construct such models to compute the time-development of the concentration field, and there is no computational problem in specifying an arbitrary spatial distribution and time-sequence of sources. The computational effort for all this may be little more than required to evaluate solutions of one of the more complicated analytic models.

There is an important distinction between numerical models in which the flow field is either steady or specified, and those in which the flow field (and possibly also mixing parameters) are computed as a function of time. Models of the former type may be classed as dispersion (diffusion/advection) models. Those of the latter type are <u>also</u> dynamical models: i.e. they have to represent the dynamics of ocean circulation in order to compute the flow as they proceed. Dispersion models are substantially simpler and easier to construct: mathematically the problem they solve is linear, whereas the dynamical problem is non-linear, and the methods required are still to some extent a matter of debate. As a rough guide, a dynamical/dispersion model is probably an order of magnitude more complex than a dispersion model of the same spatial resolution.

A further and substantial advantage of the dispersion model is that there is no particular difficulty in representing the dispersion processes in an isopycnal co-ordinate system: this is believed to be very important in the ocean, which is stably stratified so that dispersion is a highly anisotropic process. The construction of dynamical models in isopycnal co-ordinates is still in its infancy (see, for example, Black and Boudra, 1981). It is also easy to include the effects of scavenging, biological processes, sediments, etc., in diffusion models - the only limitation being the ability to find an adequate parametrization.

Some dispersion models exist, although they are not very common since physical oceanographers have primarily been interested in models of ocean dynamics. Examples are the two-dimensional horizontal model of Kuo and Veronis (1973), the two-dimensional meridional isopycnal model of Shepherd (1980), and the three-dimensional Cartesian model of Fiadeiro and Craig (1978). In all of these, the flow field is specified somewhat arbitrarily, but there should be no fundamental difficulty in constructing a three-dimensional model similar to that of Fiadeiro and Craig with a flow field determined by real ocean data: for example by using consolidated output from one of the more elaborate 3-D dynamical models, or by using inverse methods as suggested by Wunsch (1981).

Examples of the use of models of this level of complexity for waste disposal calculations are the application of a box model by Webb and Grimwood (1976) for a radiological safety assessment, and the calculations reported in Appendix VIII of this report. Medium-elaboration box models in an isopycnal frame of reference are currently being developed for deep-sea waste disposal and assessment, and for the evaluation of the CO_2 problem.

The principal difficulty in such models is to arrive at an acceptable specification of transport and mixing coefficients. There is a lack of independent data sets with which to check on the postulated exchange coefficients, and the ability of the models to predict contaminant transfer is therefore questionable. As in simpler models (or some more elaborate ones) the reliability of medium-elaboration box models depends strongly on the reliability of the parametrization of processes not modelled in detail.

4.2.9. Complex numerical models

More complex models are necessary if more realistic representation of the ocean system, or dynamical calculations, are required. Such models may contain thousands of 'boxes' and are usually of finite-difference type.

It is not yet clear whether such models offer any practical advantage over the simple methods outlined above. They are so far not available in isopycnal co-ordinate systems (see also Section 4.2.8) and the impression of realism conferred by recognizable coastlines and topography may be somewhat spurious. They would represent, however, the application of 'state of the art' methods to the problem and some usage may be required to determine whether the increased complexity is necessary.

So far models in this class have been used mostly for physical oceanographic dynamical calculations. The necessary computer code for diffusion/advection calculations is generally available (to allow for transport of heat and salt), however, and some tracer work has been carried out by Sarmiento and Bryan (1982). There is usually no purpose in recalculating flow fields for every diffusion/advection calculation (it would be an extreme contamination scenario indeed which was capable of perturbing the ocean circulation). Generally flow 'histories' may be created, and used as the basis for subsequent dispersion calculations (this does not apply to the use of time-dependent eddy-resolving models, where simultaneous development of flow and tracer distribution is of the essence).

For some purposes, a simple diagnostic calculation (for example, Holland and Hirschman, 1972) where the flow is inferred from the observed density distribution, may be adequate, and work along these lines is in progress. The results of such calculations however sometimes have bizarre features, such as unreasonably large upward flows (Veronis, 1975) and it may be preferable to use more costly prognostic calculations, in which the density field is also computed from given boundary conditions (heating, cooling, wind, etc.) The model developed by Bryan and his co-workers (Semtner, 1974) has become a de facto world standard for such calculations, and developments of this for waste disposal calculations are in progress. A set of nested, coupled, high-resolution models for a release site is also being developed in the USA. Various biological and geochemical processes of the appropriate scale are being included in the models.

Finally, there are eddy-resolving general circulation models (for example, Schmitz and Holland, 1982, and Robinson et al., 1979) capable of showing the statistical effects of eddies on tracer distributions. These are at least one order of magnitude more costly and complex than the prognostic models of the same vertical resolution, and require the largest and most powerful computers in existence. Development of such models for dispersion calculations is also under way.

The high-resolution models discussed in this Section are certainly one tool which may be used for estimates of contaminant transfer but they contain uncertainties as great as many simpler models regarding realistic long-term transports. None of these models is known to maintain the observed oceanic density field on long time-scales and thus may have trouble representing such things as slow diapycnal mixing. Similarly, problems remain as to the location of deep vertical upwelling in these models. The main role of high-resolution models may be in 'process studies' to test the parametrization used in models of lower resolution.

4.3.   Conclusions

There is no single model appropriate for all purposes, since the purpose has as much bearing on the appropriateness of a model as the processes it represents. It is seldom necessary to make a detailed simulation of a process in order to make an adequate estimate of its salient properties. Each of the models outlined above is potentially useful for certain calculations relevant to estimates of the dispersal of contaminants. The simpler ones are in general useful for special cases, for order-of-magnitude estimates, as aids in the validation of their more sophisticated relatives, and may be all that is required in some cases. An illustration of how several simple models may be linked, and applied to waste dispersal, is to be found in the work of Kupferman and Moore (1981).

If one wishes to carry out detailed estimates for a wide range of contaminants, including all the principal processes, some sort of numerical model may be appropriate. In order to keep the complexity and difficulty of interpretation and comprehension to a minimum, this should be the simplest that will do the job. Increased complexity is no guarantee of increased reliability. Simplicity is also an advantage when the results have to be fed in to other computations which are complex in their own right, where the ocean dispersion calculation has to be regarded effectively as a subroutine from which results can be obtained whenever required.

It is not yet clear just what level of complexity will be needed. This clearly depends on, among other things, the need for accuracy in any particular hazard situation. The techniques for selecting appropriate models are discussed in the following Section.

5. RECOMMENDED MODELS

In this Section we make fairly specific recommendations on the types of model that should be used to calculate contaminant concentration arising in the ocean due to a localized source on the sea floor. Our emphasis will be on simple, easily applied models, several of which were devised by the Working Group. Although these are still being developed and their consequences explored they already appear to be adequate for use in the appropriate circumstances. The development and use of simple numerical models that do not yet exist but which are anticipated to be developed easily and then used operationally are also recommended. Nonetheless, there is always the possibility that these models will lead to unforeseen complications or difficulties and our recommendations should be taken as guidance rather than infallible truth. Continued interaction is required between those using the simple models that we propose and those developing the more complicated models that may be necessary in some circumstances.

5.1.   Processes to be modelled

The Working Group has concluded that in the construction of models the following processes should be considered:

(a) the movement and mixing of water within the ocean basin: although this takes place preferentially along isopycnal surfaces, the relatively slow vertical (or diapycnal) motions are of special importance;

(b) the radioactive decay or chemical degradation of contaminants;

(c) the interaction of contaminants with particulate materials of various sorts, including biogenic particles, both within the water column and on the bottom; and

(d) mixing (e.g. bioturbation) and diffusion into (or out of) surficial sediments.

It is unlikely that all of these will be important for any single contaminant. The range of contaminant properties is however so great that different processes are dominant in different cases. Models are required that are capable of handling most combinations of processes, possibly all simultaneously.

The representations of the processes concerned need not necessarily be, and generally cannot be, detailed simulations of them. Simpler representations (parametrizations) may be required. Judiciously applied, these are often adequate for practical purposes. However, a parametrization should not be used beyond its limits of validity, and it must be remembered that features of the process which have been neglected may sometimes be important. It may be possible to use an improved model, but sometimes a lack of understanding of the real process (e.g. diapycnal mixing) may render the results of any model imprecise.

5.2. Definition of domains

When specifying the oceanic concentration of a contaminant dispersing from a point source on the sea floor it is convenient to consider three different regions: the extreme near field, the near field, and the far field. These are defined (rather loosely) as follows:

(1)   The extreme near field is that part of the benthic boundary layer in the vicinity of a release; it is essentially the region where the formula for the concentration in the near field breaks down due to the difference between the mixing rates in the benthic boundary layer and in the overlying water. In practice it might be of the order of 100 m thick and perhaps 30 km in radius.

(2)   The near field is the region in the vicinity of the release in which the concentration is significantly greater than (say, more than double) the ocean, or ocean basin, average. Its size is variable but usually less than about 10% of the volume of an ocean basin, and may be very much less for very long-lived, or very short-lived contaminants. The average concentration over a smaller region within the near field may be required for some purposes (see Appendix III).

(3)   The far field consists of the rest of the ocean.

These concepts can be illustrated by reference to the solution of the Simple Finite Ocean Diffusive Model described in Appendix IV. Near the source the solution behaves like that for a point source in an infinite ocean, with the concentration given by:

$$C = (Q/2\pi) \; [(K_H K_V) \; (x^2 + y^2) + K_H^2 z^2)]^{-\frac{1}{2}} \tag{5.1}$$

and depends on the diffusivities K_H and K_V. At least the latter, K_V, can be expected to be much higher in the benthic boundary layer than above it. If d denotes the thickness of the benthic boundary layer then it is appropriate to use Equation (5.1) within an extreme near field of height d and radius $(K_H/K_V)^{\frac{1}{2}} d$, with K_H, K_V in Equation (5.1) and in this definition of the radius of the extreme near field equal to the values appropriate to the benthic boundary layer.

In practice d might be of the order of 100 m, with K_H, K_V having values of the order of 10^3 m^2 s^{-1} and 10^{-2} m^2 s^{-1}, so that the radius of the extreme near field is of the order of 30 kilometres.

The near-field solution of the Simple Finite Ocean Diffusive Model is essentially given by Equation (5.1) with the diffusivities K_H, K_V having values appropriate for the main body of the ocean. The extent of validity of this has been estimated in Appendix IV, as $\lambda V(2\pi K_H)^{-1}$ vertically and $\lambda V(2\pi)^{-1}(K_H K_V)^{-\frac{1}{2}}$ laterally for a contaminant with a long half-life compared with the ocean mixing time (the far-field solution in this case is simply $Q(\lambda V)^{-1}$). These scales are of the order of hundreds of metres vertically and hundreds of kilometres horizontally for contaminants with residence times of the order of ten thousand years.

This discussion should serve to illustrate the need to discuss separately the concentration in three different regions, and suggest how one might separate the extreme near field, near field and far field. However, it is certainly not suggested that the Simple Finite Ocean Diffusive Model is always adequate for the calculation of concentrations. In the following Section we discuss more completely the choice of model for estimation of the concentration in the various regions, for contaminants with a variety of properties.

5.3.   General results affecting model selection

In the course of its deliberations the Working Group reached various general conclusions that assisted in the selection of appropriate models and parametrizations, and aided constructive thought. They are useful for making initial estimates (e.g. of the relative importance of various processes) and are summarized below. While thought to be valid in the light of present knowledge, they should be kept under review. This is especially true for those conclusions that depend on the value chosen for various parameters. Cases where they fail will draw attention to areas where further work will be required.

(1)   Within the assumption of first-order scavenging, contaminants are likely to be well equilibrated between particulate and water phases in the ocean interior, i.e. outside relatively thin transition layers close to boundaries, sources and sinks (Appendix IX). The effects of these transition layers appear to be relatively minor.

(2)   For steady-state problems the effects of scavenging and/or mobilization at boundaries may be adequately parametrized using deposition velocities. The importance of interior scavenging of the ocean water inventory may often also be adequately estimated by this method, but the details of the interior concentration field may not. In any case, the specification of the appropriate deposition velocity demands careful attention to the processes at work. This parametrization is usually inadequate for time-dependent problems, except in special cases (Section 3.2.2 and Appendix V).

(3)   Within a radius of the order of 10-100 km around a source, the effects of advection on the concentration, averaged over times

of about 1 year or more are small, and diffusive solutions are adequate (Appendix IV).

(4)   In the deep ocean, the minimum area for spatial averaging required for models of transfer into food chains is of the order of 100 km^2, and the minimum period for time-averaging is 1 year or more (Section 2.1 and Appendix III). Ensemble-average concentrations (calculated using eddy diffusion coefficients, for example) are therefore adequate for most purposes (Section 2.4).

(5)   The concentration field in the extreme near field is only of relevance when considering effects on abyssal organisms (Section 3.3.3).

(6)   Fluctuations are not likely to exceed the expected (average) concentration except in regions of high concentration gradient (e.g. near the source) (Sections 2.4 and 3.1.3).

(7)   The characteristic time for the development of the expected concentration field near the source (say within 300 km radius) is of the order of a few years, and the concentration field is therefore in approximate steady state with the recent average release rate over such a time period.

(8)   Within a reasonably general class of models, the laterally averaged concentration field is given by the solution of a one-dimensional model (i.e. the ratio of contaminant inventory in the sediments to water inventory is no greater, or less, for dispersion in a three-dimensional ocean than is given by a one-dimensional model) (Appendix VII). Since the detailed lateral distribution does not affect the inventories, one-dimensional models have wider applicability than might have been expected.

(9)   The detailed behaviour of the bottom boundary layer is unimportant except insofar as it affects average vertical transfer (i.e. K_V due to boundary mixing), or affects the concentration in the extreme near field.

(10)   Transports by organisms, other than those associated with downward transport in biogenic particles, and mixing of sediments, do not significantly affect the concentration field established by physical and geochemical processes (Appendix III).

5.4.   Model selection

The choice of an appropriate model for a particular contaminant will depend on its natural lifetime (due to decay or degradation, on its reactivity with various forms of particulate material and on the purpose of the calculation. Suitable models for use in time-dependent and steady-state problems and for the near field and far field are listed in Table III. For the extreme near field the point-source solution of Appendix IV, modified for finite source size according to the discussion in Appendix VII, would be appropriate.

TABLE III

RECOMMENDED MODELS

Near field	(a)	Simple Finite Ocean Diffusive Model (Appendix IV)
	(b)	Modified for finite source size and scavenging (Appendix VII)
	(c)	Plume solutions if the size of the near field exceeds the scale K_H/U within which diffusion dominates (Appendix IV)
Far field	(a)	Well-mixed box (for contaminants with a long residence time)
	(b)	The one-dimensional scavenging models of Appendices VI and IX
	(c)	The Simple 3-D Diffusive Model with Scavenging (Appendix VII)
	(d)	A medium-resolution box model
	(e)	Finite-difference models in 2- or 3-D

Notes: 1. At least the processes listed in Section 5.1 must be included where appropriate.

2. The choice of model may be influenced by the need to compute concentrations relevant to a particular food chain or exposure pathway.

It will usually be sufficient to use the simpler models within each section of Table III. If the resulting prediction is very sensitive to some poorly known oceanic parameter (such as K_V) there may be no advantage to be gained from a more complicated model that allows for a more detailed representation of some other less important process (such as the depth dependence of interior scavenging, for example). However, there may be circumstances under which the use of a more elaborate representation is convenient or enables one to test the importance of extra features.

Table III has been subdivided into near field and far field. Under some circumstances it may be necessary to allow for the existence of different ocean basins.

We have only listed in Table III the recommended models for use in the steady-state problem. For cases of practical interest the solution of a time-dependent problem would not differ significantly from the steady-state solution in the near field. In the far field, time-dependent versions of the steady-state models may be required and their solution will generally require the use of numerical techniques even if the steady-state model is analytical.

5.5. Model sensitivity and reliability

As has been emphasized throughout this report, a model's ability to predict accurately the concentrations arising from the release of a contaminant depends on whether the model provides a realistic representation of all the important processes affecting the contaminants' dispersal. This is usually difficult to determine, especially quantitatively.

It is thus recommended that so far as possible models should be run over the full range of possible variability of their input parameters and that the range of predicted concentrations be presented as part of the output. The sensitivity of the prediction to the input parameters so obtained provides valuable information about the model's performance. In particular:

(a)   for those cases where a "best estimate" of the concentration field is needed, it indicates the range of possible uncertainty about this estimate;

(b)   for those cases where maximizing transport concentration fields are needed, it enables the determination of the maximum realistic concentration field within the model's assumptions and indicates its difference from the "best estimate"; and

(c)   for all cases, it indicates the parameters to which the model's results are most sensitive. This can lead to an evaluation as to whether this is due to the model's parametrization of known important processes or to a lack of knowledge of these processes on which to quantify better their effects.

Models should always be evaluated in terms of the sensitivity of their results to the importance of neglected processes. In some cases it will be possible to do this on the basis of the results of simple models and will not necessitate the detailed solution of a much more complex model. This approach has often been taken in this report in examining the significance of various processes.

Lastly, models should wherever possible be checked against existing data sets for the distribution of natural and/or anthropogenic substances. It must be remembered however that multi-parameter models may reproduce the distribution of substances, especially if the parameters have been adjusted for that purpose, but may nevertheless be unreliable for the prediction of the concentration field of a substance that is introduced or removed in some different manner. Thus, the choice of data sets for model "tuning" or verification must be done with considerable care (see for example the results of Appendix IX).

41

6. FUTURE RESEARCH NEEDS

The models recommended in the previous Sections have been chosen both on the basis of the specific needs for deep-sea waste disposal and on our knowledge of the processes controlling the transport of contaminants in the ocean. For the most part this has resulted in relatively simple existing models being selected or new ones formulated.

Progress in the future seems to depend primarily on obtaining increased knowledge of critical processes and their parametrization along with the concurrent development of slightly more elaborate models for the purpose of testing the importance of various processes or assumptions. Both aspects are discussed below.

The continental slopes are regions where much of the vertical exchange by physical processes is thought to occur, and also where food chains leading to deep water may be located. They are also the places where many isopycnal surfaces contact the sediments. It is therefore appropriate for them to be given emphasis in the research activities outlined below.

6.1. Processes needing research

6.1.1. The geochemical interactions in the water column

The modelling attempted by the Working Group has made use of first-order kinetics, partly suggested by the distribution of naturally occurring thorium isotopes in the ocean, to parametrize scavenging onto fine particles. Additional mechanistic and kinetic studies are needed to establish how contaminants interact with particle surfaces and the rates and reaction order of the interactions. Recognizing that suspended particles include aluminosilicate, biogenic and authigenic particles, the chemical and physical nature of particle surfaces needs to be considered. Such studies would include characterization of the formation and decomposition rates of organic matter, the dissolution rate of tests and the rate of biological processing of particles and formation of faecal pellets. Since contaminants associated with particles are eventually removed from the water column by settling, investigation on the geographical variations of the flux and composition of particulate matter are also needed. Contaminant uptake by plankton and faecal pellet formation needs to be related to nutrient concentrations and productivity.

6.1.2. Vertical mixing and the thermohaline circulation of the oceans

All the models discussed are sensitive to the magnitude and spatial and temporal dependence of vertical (or diapycnal) exchange processes by both mixing and water movement. The nature of the thermohaline circulation and diapycnal mixing processes which maintain the isopycnal topography of the oceans are still not well described or understood, and continued efforts to solve these classical problems of physical oceanography are required. Since the circulation must ultimately depend on atmospheric conditions (and vice versa), elucidation of the changes occurring on decadal and longer time-scales is also desirable.

The specific requirements regarding information on vertical exchanges for contaminant transport are similar to those of climate research or programmes to determine the generation mechanisms for the distributions of natural oceanic properties (for example, temperature and salinity).

6.1.3. Horizontal mixing and advection

Extension of existing models to give realistic horizontal distributions requires increased information on the large-scale oceanic horizontal circulation and horizontal (or along isopycnal) mixing. Of particular importance is the determination of the rates of exchange between ocean basins or regimes and the time-scale for horizontal mixing within them.

Much of the current oceanographic research on mesoscale processes (eddies, rings, etc.) using current metres, long-range float tracking, the interpretation of natural and anthropogenic tracer distributions, together with extensions of the classical dynamical method is aimed at exactly these questions and needs to be maintained. Novel methods of attacking the problems should, of course, be sought and supported.

6.1.4. Biogeochemical processes in sediments

A contaminant associated with particles may be subject to various transformations (e.g. dissolution, microbial oxidation) in the surface sediments after its deposition. These transformations can result in release of contaminants to sediment pore waters and will vary as a function of the flux of organic matter to the bottom and its location in the ocean. Since they need further clarification:

(1)   Studies are needed of interaction between particles and contaminants in the sediment/pore-water system. These studies could, in many respects, use the same approaches and tools as similar studies in the water column (Section 6.2.1) and should include research (via tracers or analogues) on possible speciation and oxidation state transformations of contaminants likely to be deposited in sediments.

(2)   Further work is needed on the geographical variation of oxidation-reduction conditions within the depositional environment.

(3)   Studies of interactions between the benthic organisms and sediments are needed to characterize the effect of the fauna on chemical transformations in the sediments, accumulation of contaminants in organisms and the effects of the fauna on the flux of contaminants across the sediment/water interface.

(4)   The uptake of tracers by bottom sediments is affected by the rate of particle mixing by the benthic fauna. This should be determined in a variety of environments by consideration of the distribution of tracers such as microtektites, volcanic glass and natural and artificial radionuclides.

6.1.5. Oceanic distributions of geochemical tracers

Tracers exist that have a variety of sources and sinks and are both particle-reactive and soluble. Their distributions thus provide a test of a model's ability to represent geochemical interactions and physical processes. This however requires knowledge of the distribution on a scale compatible with the resolution of the model used. In general, further work is needed to bring tracer databases to a useful level, at least in terms of basin-wide variations.

Advantage should also be taken of several ongoing oceanic tracer "experiments". These include the development of transients added to the surface oceans as a result of atmospheric weapons testing and industrial activities. Particle-reactive transients include the artificial radioisotopes of plutonium and caesium, as well as industrially introduced stable lead. The measurement of distributions of these tracers on a time-scale of years to decades provides valuable information on transport rates.

Although the cycling of many tracers may result in high concentrations at the bottom, those with natural bottom sources are particularly valuable for studying mixing and geochemical interactions in a situation perhaps more closely resembling that of a release from waste disposal. The vents associated with hydrothermal circulation through the oceanic ridge system offer one such opportunity. Tracers of potential use include ^{3}He, ^{210}Pb and manganese.

6.1.6. Quantification of biological processes

The Working Group attempted to quantify the effects of deep-sea animals. More information is however needed to further this aim. In particular:

(1)   Knowledge of the extent of horizontal and vertical migrations of adult deep-sea demersal fish is required to assess the spatial scales of contaminant transfer into food chains that may constitute critical pathways to man.

(2)   The distribution of biomass and species assemblages should be investigated with those food web interactions among deep-sea organisms that might transfer contaminants from a central oceanic region to peripheral upper slope regions.

(3)   Information is necessary on the fate and relative rates of supply of organic and inorganic material by both large and small particles to deep-sea communities, so that the amount remaining after respiration for biomass production and potential contaminant transfer and the amount lost to burial may both be evaluated.

(4)   The response of deep-sea organisms to particular contaminants requires investigation to assess (i) their accumulation and metabolism (especially concentration factors) and (ii) effects on species composition at the community level. Information on colonization of sessile organisms on solid surfaces placed in the deep sea might be used to evaluate the potential impact of physical alterations at disposal sites where mobile animals may be caused to aggregate in response to a concentrated food supply.

6.2.   Models for research purposes

More elaborate models than those discussed in this report can serve as useful research tools for investigating the importance of various processes and the limitations of parametrizations used in simpler models. Their evolution into useful operational models should always depend on their increased complexity being supported by adequate knowledge of the mechanisms involved.

The Working Group has identified a number of useful models but others certainly exist. An important elaboration seems to be the introduction of vertical or horizontal variations to various basic models. Thus, one should consider:

(a) the construction of three-dimensional models with simple flow fields and particulate scavenging processes: such a model could be the elaboration of the 3-D model of Fiadeiro and Craig (1978);

(b) the inclusion of vertical variation in various of the physical parameters in one-dimensional models such as are discussed in Appendices VI and IX;

(c) the inclusion of a horizontal and vertical variation in the flow and scavenging regimes in the hybrid model of Appendix IX to analyse the nature of particle/water and sediment/water interactions: such a model could also be extended to include time-dependent contaminant releases;

(d) the use of eddy-resolving or stochastic models to investigate the importance and nature of fluctuations about the average concentration and help resolve questions about the evolution to the steady state of the concentration field of time-dependent contaminant release; and

(e) the development of more realistic oceanic plume models to investigate concentration fields in the near field.

Appendix I

GEOCHEMICAL PROPERTIES AND OBSERVATIONS OF THE DEEP OCEAN

1. PARTICLE DISTRIBUTIONS AND FLUXES

Though particles are sparse in sea water, they control the oceanic distributions of trace metals such as Th, Pu, Pb, Fe, Mn, and Cu. The particle concentration of ambient ocean water is only about 1 g in 100 t of sea water and is composed of biogenic carbonates, silicates, organics, and terrigenous detritus. Inorganic precipitates such as oxides of iron and manganese do not significantly contribute in bulk weight but could be important for the scavenging of elements. The relative composition of the particle components varies depending on location and depth, so that if they have different adsorption characteristics for different elements, different scavenging regimes are expected in different oceanic sites.

1.1. Concentrations of suspended particulate matter

The large number of samples for suspended particles taken during the GEOSECS programme made it possible to map the particulate matter concentration in the north-south section of the Atlantic (Figure A.I-1). The mapping may be extended to the Pacific in the future.

As shown in Figure A.I-1, the distribution of particulate material with depth in the oceans is characterized by (a) high concentrations in surface waters decreasing across the thermocline, (b) low concentrations, $\sim 10\ \mu g \cdot kg^{-1}$ at mid-depths, and (c) increasing concentrations toward the bottom. Since the surface enrichment in particles is caused by the biological productivity, the concentrations vary with location. The near-bottom region of high particle concentrations, referred to as the nepheloid layer, is composed predominantly of inorganic particles (i.e. resuspended bottom sediments) and again is highly variable, depending on bottom currents and availability of particles. While variable in particle concentration, the nepheloid layer occurs in many places of the Atlantic, for example (see Figure A.I-1). The mid-depth maximum in particle concentration can be explained by advection of resuspended shelf or slope sediments, or by detachment of bottom boundary layers with associated suspended sediment.

1.2. The vertical flux of particles

Direct observation of the vertical flux of particles at various depth horizons in the oceanic water column has been made by sediment trap experiments. Although they are rendered complex by ambiguities concerning trapping efficiencies and the seasonality of particulate fluxes, the data obtained by the PARFLUX group at various oceanic sites are summarized in Figure A.I-2 from Honjo et al. (1982).

In the open ocean away from continents, e.g. the North Pacific and the Sargasso Sea, the settling fluxes of particles are 10-20 mg m^{-2} d^{-1}, with some variation with depth. In the subtropical Atlantic, the material fluxes are approximately constant with depth at

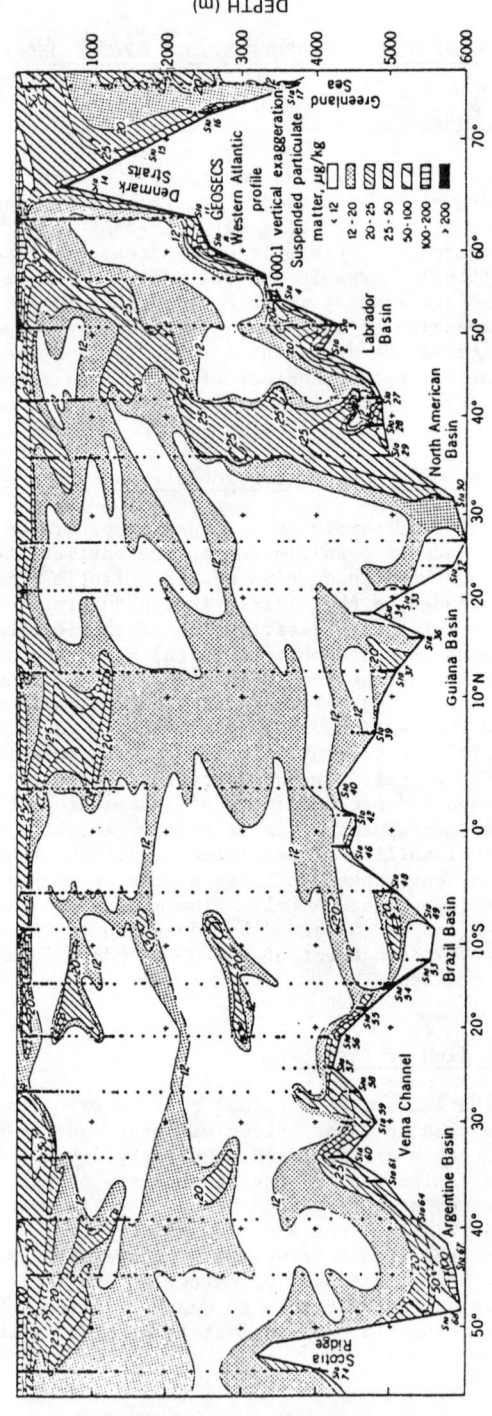

FIG. A.I-1. Isocontours for concentrations of total suspended matter in Atlantic waters measured by the Woods Hole and Lamont-Doherty groups (from Brewer et al., 1976).

about 50 mg m^{-2} d^{-1}. In the Panama Basin, the flux increases from 110 mg m^{-2} d^{-1} at 700 m to 180 mg m^{-2} d^{-1} at 3800 m. The general trend of increase in particulate flux from open ocean regions toward the margins of the ocean basin is consistent with the variation of input of terrigenous materials and biological productivity.

The sediment trap experiments indicate that the fine detrital clay particles always increase with depth both in relative content and in the vertical flux. This may be due to a horizontal source of fine clay particles from continental margins or elsewhere which is converted to the vertical flux, or may reflect resuspension from the bottom. However, it is not well known with what efficiencies fine particles are trapped by the intercepting cone, hence the arguments based on the fluxes may be hazardous at present. The fine particles contain "typical detrital elements" such as Al, Ti, K, Th and rare-earth elements whose composition is not significantly altered during settling through the water column (Brewer et al., 1980).

The ambiguities in the variation of trapping efficiencies are less for the large, fast settling, biogenic particles. As shown in Figure A.I-2b, the flux of organic matter decreases with depth, as expected from bacterial oxidation in the water column. The major change occurs in the upper 1000 metres: below that depth, the organic carbon flux decreases only slightly. The hard skeletal particles, carbonates and silicates show either a slight decrease or are almost constant with depth in their vertical fluxes (Figures A.I-2c and 2d), suggesting that the dissolution of these materials is relatively small within the water column. These biogenic particles are the conveyor of "biogeochemical elements" whose distributions are generally characterized by depletion in the upper layer and enrichment in the deep layers (see the next Section). The surface depletion is due to biological uptake and the deep enrichment is a result of oxidation and dissolution of the carrier phases.

2. DISTRIBUTIONS OF TRACE AND RADIOACTIVE ELEMENTS

The purpose of this Section is to illustrate the known behaviour of a variety of trace and radioactive elements in the deep sea. The various species are categorized in terms of the combination of source, transport and sink phenomena which determine their distributions. For simplicity we do not consider the physical phenomena which are, to some extent, common to all. Rather we focus on the variety of chemical processes in the hope that these can be parametrized in the general dispersal model.

With the important recent advances in analytical capability, primarily due to new technology, valid data for a variety of trace and radioactive elements are being produced at an increasing rate. This is an empirical activity, the choice of particular species often being dictated by capability rather than a priori determination of significance. Mechanistic and kinetic studies are only beginning. Thus while we think that for a particular element, one or a combination of processes is important in governing their distribution, we are not able in general to quantify the rates or understand their underlying causes. Radioactive production and decay are the obvious (and only) exceptions.

The important processes controlling the water column distributions for elements are listed in Table A.I-I, grouped according to the mode of input, removal, and internal biological/geochemical processes. In the

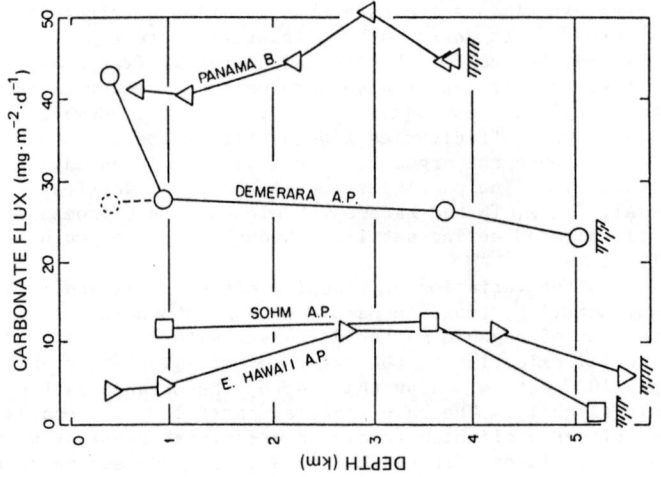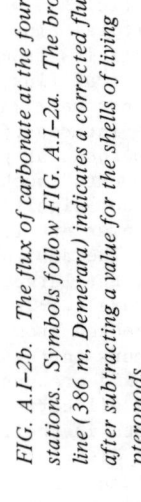

FIG. A.1-2b. The flux of carbonate at the four stations. Symbols follow FIG. A.1-2a. The broken line (386 m, Demerara) indicates a corrected flux after subtracting a value for the shells of living pteropods.

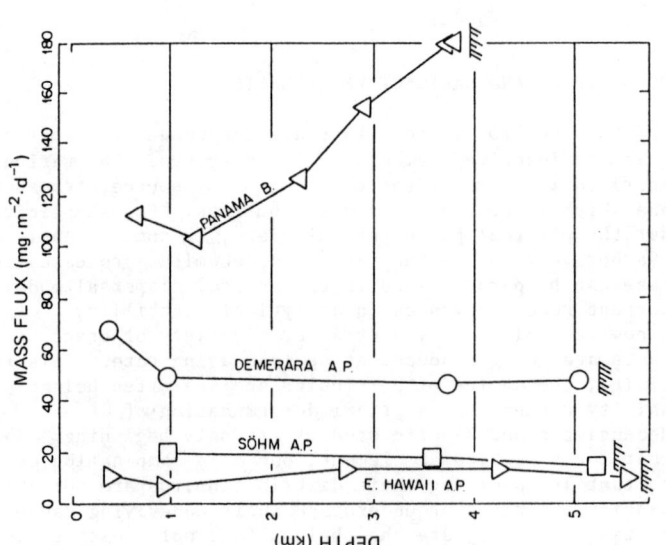

FIG. A.1-2a. The mass flux material collected at four stations by PARFLUX Mark II sediment trap with 1.5 m² opening. ▽, East Hawaii Abyssal Plain; ◻, Söhm Abyssal Plain; ○, Demerara Abyssal Plain; △, Panama Basin (Honjo et al., 1982).

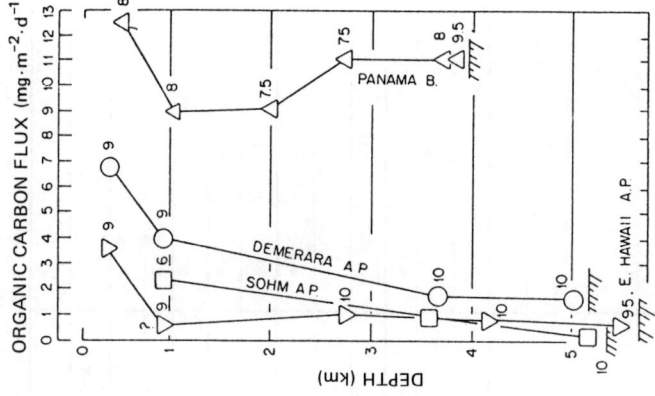

FIG. A.1-2d. The flux of organic carbon at the four stations. Symbols follow FIG. A.1-2a. Numbers on the right side of data points represent C:N ratio (by atoms).

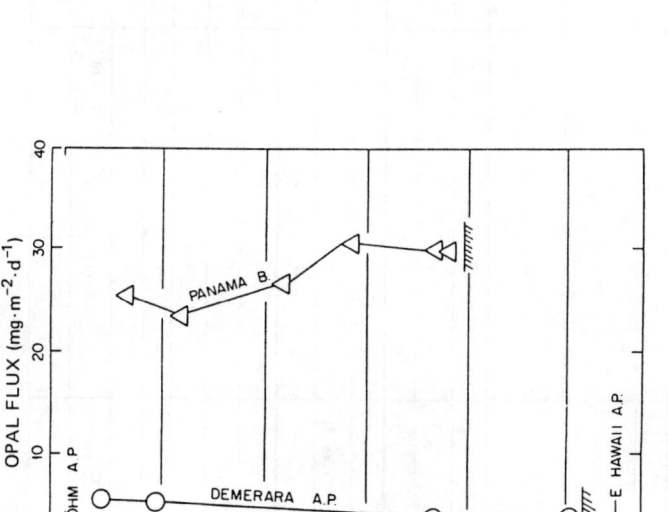

FIG. A.1-2c. The flux of opal at the four stations. Symbols follow FIG. A.1-2a.

TABLE A.I.I. LIST OF IMPORTANT GEOCHEMICAL PROCESSES

		INPUT				INTERNAL PROCESS		REMOVAL			
		Atmosphere	River	Bottom release	Radioactive production	Biological Uptake/Release	Scavenging	Sediment	Atmosphere	Radioactive decay	Consumption
I	^3H	*								*	
	^3He			*					*		
	^{222}Rn	*		*	o				o	*	
	O_2	*				o					*
	CO_2, $^{14}CO_2$	*		*		*		o			
	$228, 226Ra$			*		*				*	
II	PO_4, NO_3, Cd		*			*		*			
	Sr, ^{90}Sr	*90	*			*		*		*90	
	Se(IV)	o	*			*		*			
	Si, ^{32}Si	*32	*	*		*		*		*32	
	Zn, Ge, Cr(VI), Ba, Ni		*	*		*		*			
	Se(VI)	o	*	*		*		*			
III	Be, $^{7,10}Be$	*7,10	*	*	*	*	*10	*		*7	
	Pb, ^{210}Pb	*		*	*210		*	*		*210	
	Mn		*	o		*	*	*			
	Cu	*	*	*		*	*	*			
	$239, 240Pu$	*		o		*	*	*			
	$234, 228, 230, 232Th$		*232	o	*234,228 230	o	*	*230, 232		*228, 234	
	^{137}Cs	*				o	*	*		*	

* Process important in controlling distribution.

o Process occurs for specified element but is of relatively minor importance.

52

Table, the elements are categorically divided into three groups. The first group (Group I) of the elements includes those either unreactive in sea water or involved in biogeochemical processes to the extent that their distributions are not overwhelmingly determined by these processes. These elements and species are useful as tracers to elucidate physical oceanographic processes on time-scales dependent on the radioactive decay times and the bacterial consumption rate (e.g. for O_2).

The second group (Group II) of elements is characterized by its involvement in the biogeochemical cycle, e.g. the uptake in the euphotic zone and the release in the water column and at the sea floor. Some elements are carried with plankton soft tissues and released at relatively shallow depths, resulting in a distribution similar to that of PO_4. Others are involved in hard skeletal materials, carbonates and silicates and released predominantly at the bottom, as is the case for Si.

The third group (Group III) comprises elements which are highly reactive to the surface of particles and are generally maintained at extremely low concentrations by scavenging. Figure A.I-3a-e shows examples of the elemental distributions from which the inferences of Table A.I-I are drawn. The profiles are for the most part from the Central Pacific, where horizontal effects are overshadowed by the vertical.

3. SCAVENGING OF REACTIVE ELEMENTS

3.1. Chemical scavenging in the ocean

Scavenging of reactive species in the oceans depends fundamentally on uptake onto suspended particles. Processes by which this occurs include sorption onto particle surfaces, co-precipitation or sorption onto Fe and Mn oxides and oxyhydroxides and incorporation into planktonic tests. Transformations between different size classes of scavenging particles will affect both the ability of the particles to scavenge (e.g. by changing surface area or properties) and the rate at which the particles settle through the water column. Both will govern the rate at which a reactive nuclide[1] is removed from the water column.

The interaction of particles with dissolved reactive nuclides is diagrammatically shown in Figure A.I-4. In this simplistic conception, settling populations of large particles (arbitrarily $\geqslant 50$ μm) and small particles (<50 μm but operationally >0.45 μm) scavenge dissolved reactive nuclides. The large-particle class includes faecal pellets and planktonic tests, carcasses and molts, while the small-particle class includes detrital particles such as aluminosilicates and fragments of large particles. Fine particles in the 1-10 μm size range are characterized by a size distribution given by:

$$dN = Ar^{-b} dr$$

where N and r are the number and radius of particles, respectively, and A and b are constants (Lal, 1980). Observations indicate that A varies depending on particle abundance but b is approximately constant (~ 4)

---

[1] In this treatment, the term "nuclide" is taken to refer to any chemical species.

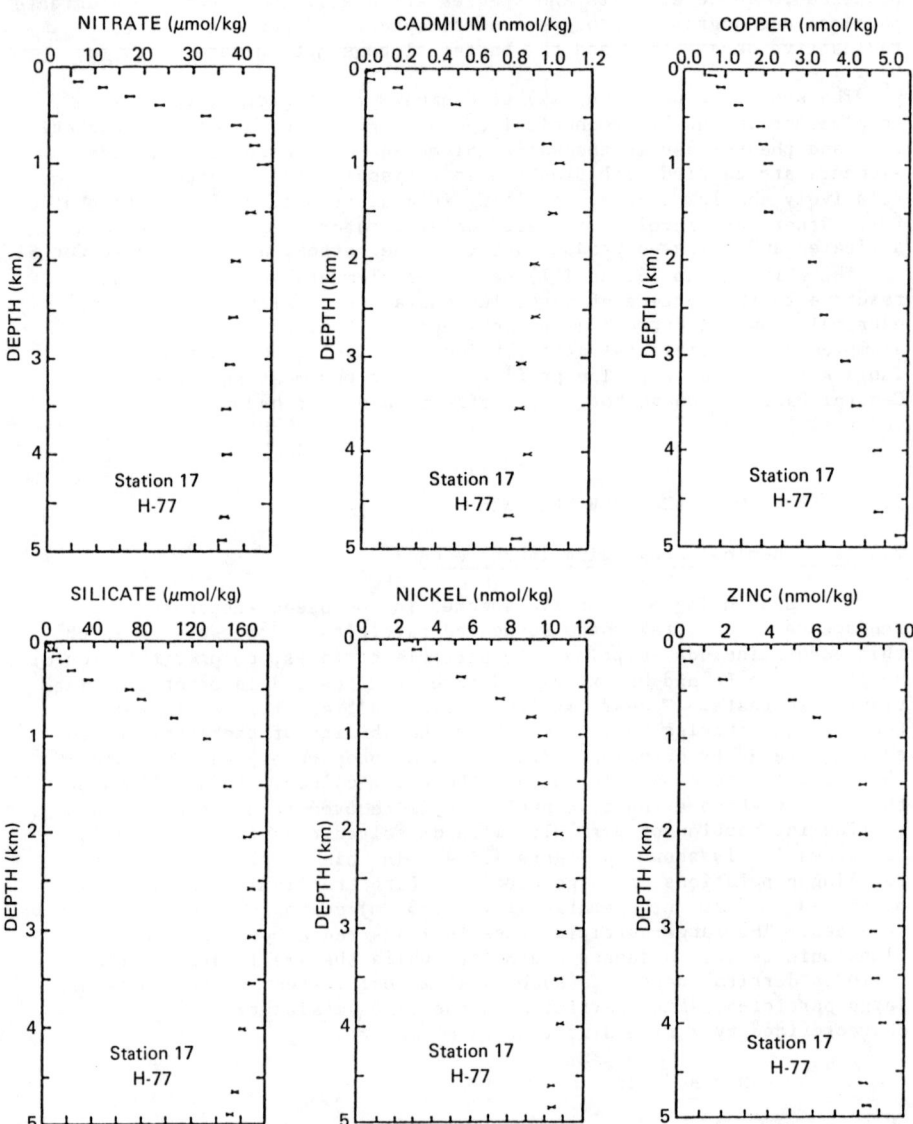

FIG. A.I–3a. *Profiles from nitrate, silicate, cadmium, nickel, copper and zinc in the eastern North Pacific (Bruland, 1980).*

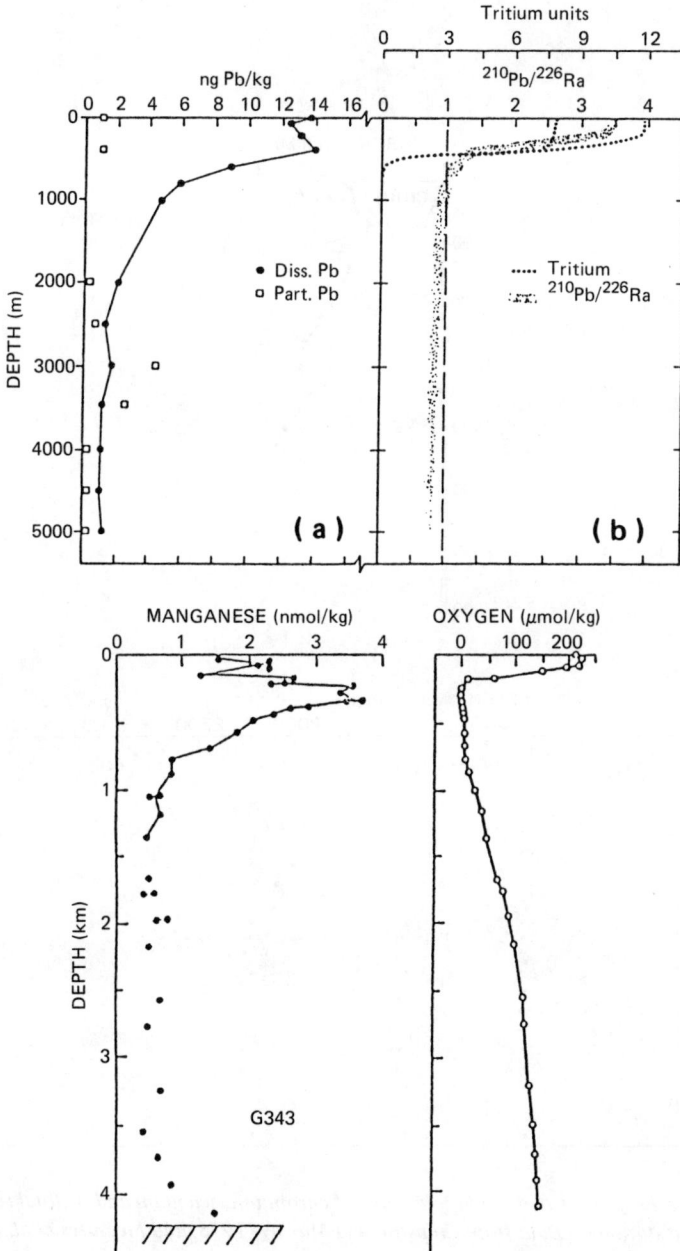

FIG. A.I–3b. Profiles of stable lead and $^{210}Pb/^{226}Ra$ (top) and manganese and oxygen (bottom) in the eastern North Pacific. (Data from Schaule and Patterson, 1980, and Klinkhammer and Bender, 1980.)

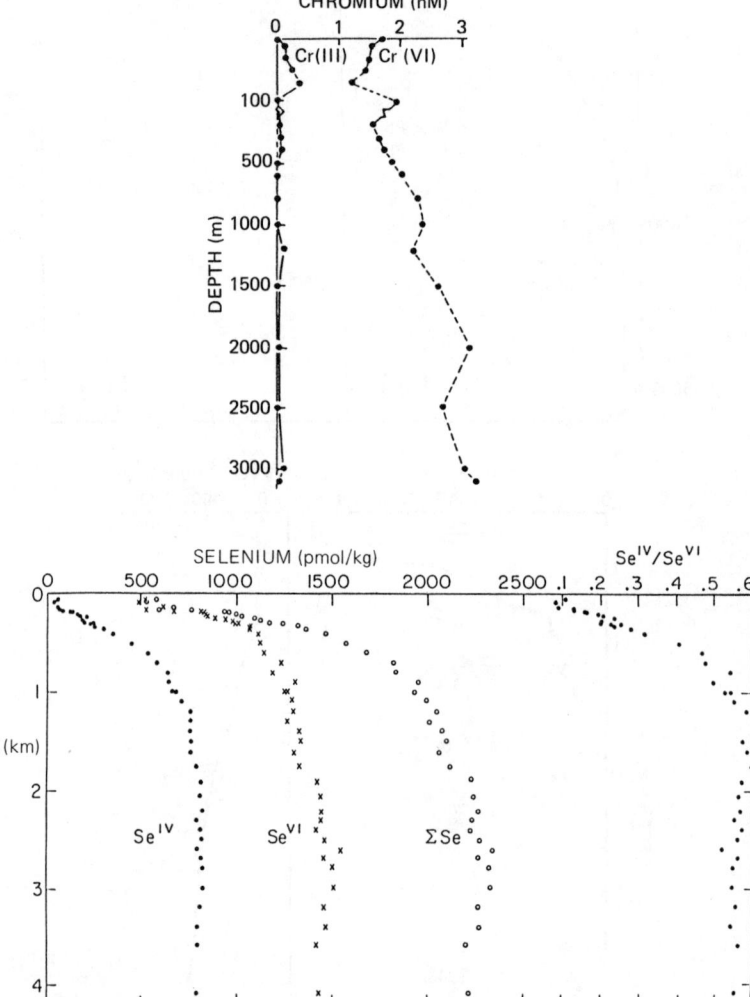

FIG. A.I–3c. Profiles of two oxidation states of chromium, selenium and of total selenium for North Pacific stations. (Data from Cranston and Murray, 1978, and Measures et al., 1980.)

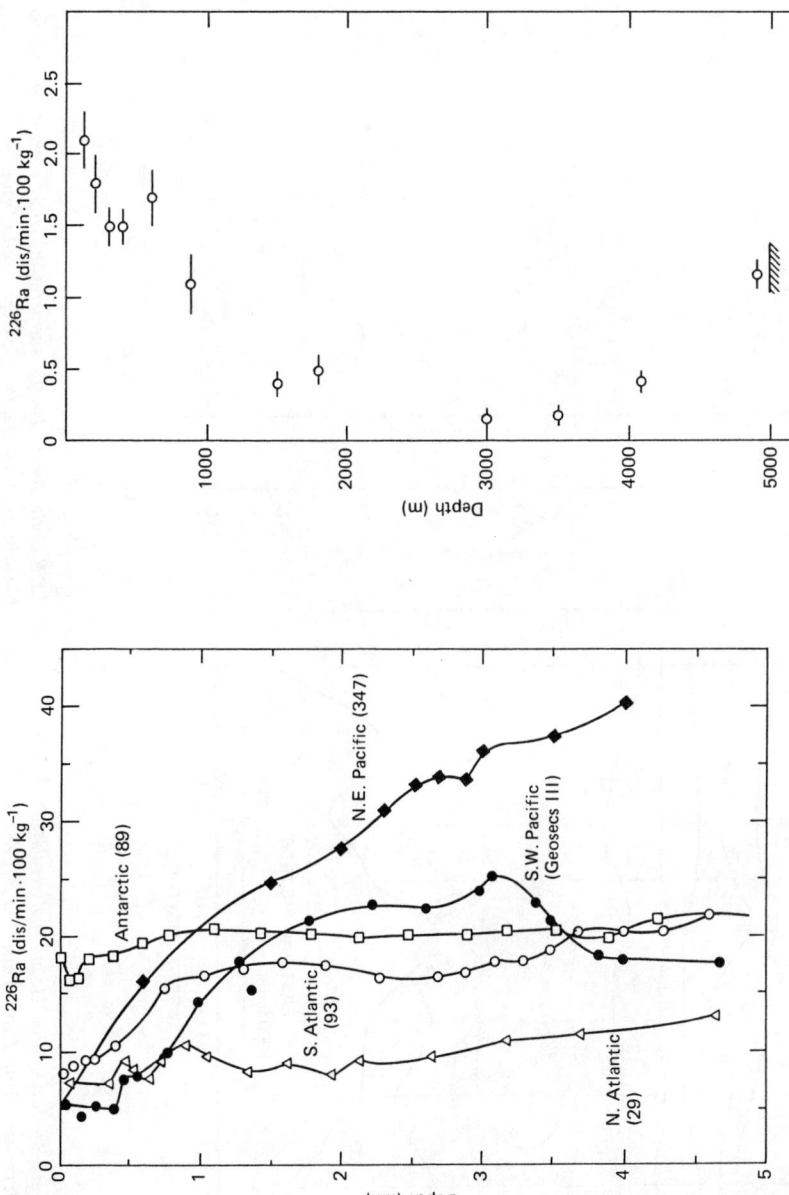

FIG. A.1-3d. Profiles of ^{226}Ra in different ocean basins (left) and ^{228}Ra in the North Atlantic (right) (Cochran, 1982).

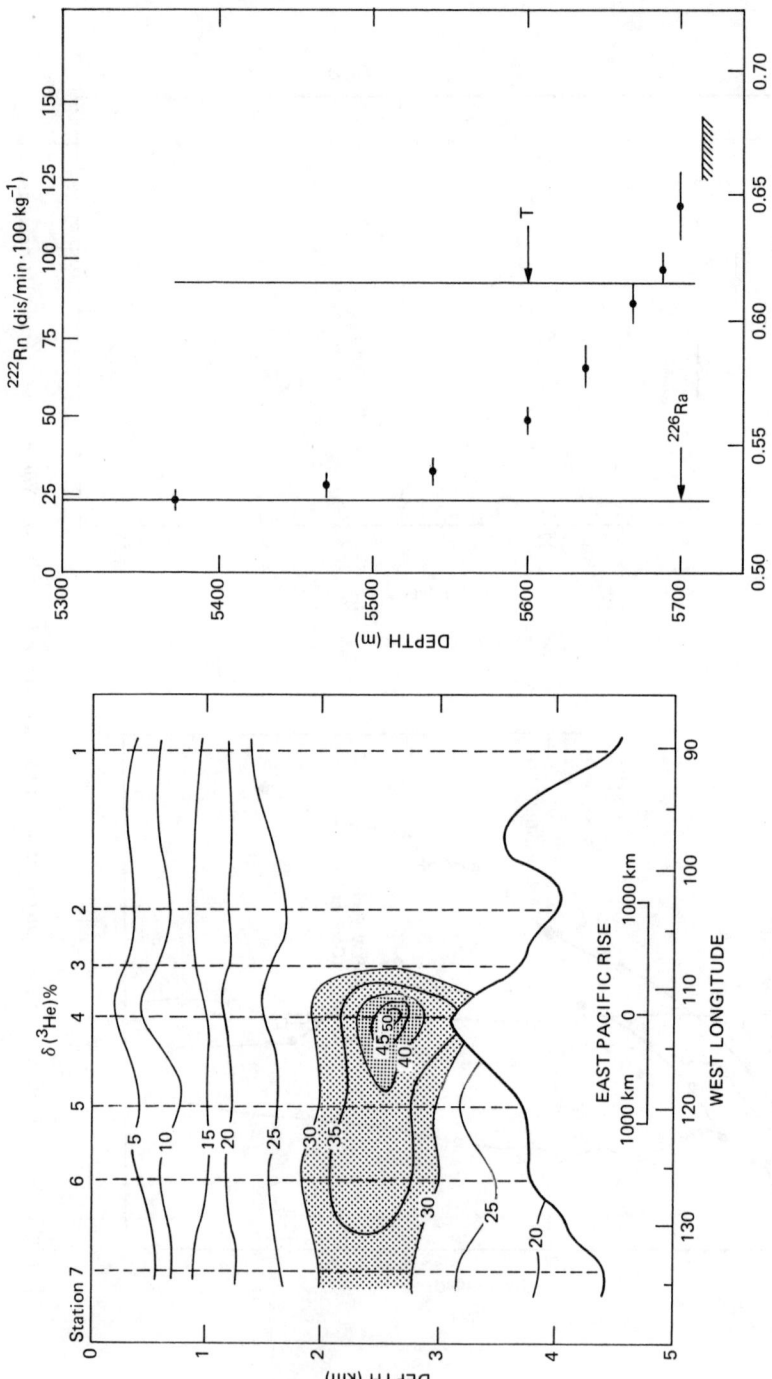

FIG. A.1-3e. Section of the ³He anomaly (relative to solubility) along 15°S in the South Pacific showing the ridge crest hydrothermal injection (left, data from Lupton and Craig, 1981) and ²²²Rn profile in bottom water of the Pacific (right, data from Sarmiento et al., 1976).

independent of location and depth. Thus, although the mass flux of
particles is dominated by the large particles, the particle surface area
is dominated by the smaller particles. With respect to scavenging and
removal of reactive nuclides from the water column, it is difficult to
ascribe the process to solely the large or small particles because of the
likelihood of constant transformations between the particle size
classes. Small particles can be incorporated into large ones by
packaging as faecal pellets or by a "sweep-out" effect (Lal, 1980).
Large particles can physically break up into smaller particles. This
disintegration may be associated with microbially mediated decomposition
of organic material which binds the large particles.

As particles sink they can both scavenge and release reactive
nuclides in the water column. This process is frequently taken to be
first order with respect to concentration of the reactive species,
although second-order uptake is possible (e.g. with PO_4 onto $FeOOH$).
Earlier work on chemical scavenging in the oceans (e.g. Craig, 1974) took
scavenging to be an irreversible uptake. More recent work with thorium
isotopes, described in some detail in Section 2 of this appendix,
indicates that a reversible process is more likely for this element.
Because first-order sorption kinetics can be readily applied to some of
the modelling efforts developed elsewhere in this report (Appendices V,
VI and IX), we describe this formulation in some detail and apply it to
the thorium data in the next Section.

With the first-order assumption, a reversible sorption equilibrium
may be written as

$$C \; \underset{k_2}{\overset{k_1}{\rightleftharpoons}} \; C_p \qquad\qquad (A.I.1)$$

where C = concentration of reactive nuclide in solution/volume
             (sea water + particles)
       C_p = concentration of reactive nuclide on particles/volume
             (sea water + particles)
       k_1 = rate constant for adsorption
       k_2 = rate constant for desorption.

The steady-state balance for a radioactive nuclide in the particulate
phase is then

$$\frac{dC_p}{dt} \; = \; 0 \; = \; k_1 \, C \; - \; (\lambda + k_2)C_p \qquad\qquad (A.I.2)$$

and

$$\frac{C_p}{C} \; = \; \frac{k_1}{\lambda + k_2} \qquad\qquad (A.I.3)$$

For long-lived radionuclides (or stable species) $k_2 \gg \lambda$, and

$$\frac{C_p}{C} \; \approx \; \frac{k_1}{k_2} \qquad\qquad (A.I.4)$$

59

DISSOLVED $\xrightarrow[K_2]{K_1}$ LARGE PARTICLES ($\geqslant 50\,\mu$m) $\xrightarrow[\text{grazing, sweep-out}]{\text{disintegration, remineralization}}$ SMALL PARTICLES ($\leqslant 50\,\mu$m) $\xrightarrow[k_2]{k_1}$ DISSOLVED

S'

s

($\geqslant 10$ m d^{-1}) ($\leqslant 1$ m d^{-1})

FIG. A.I–4. Scavenging model.

The rate constant k_1 is a function of the number of available sorption sites, which is dependent on the type, abundance, and the surface area of the particles; k_2 will depend more strongly on the chemical interaction between particle and nuclide. Thus, k_1 should be roughly proportional to the particle concentration, f, while the ratio k_1/k_2 is related to the commonly measured parameter K_d by the equation[2]

$$\frac{k_1}{k_2} = f\,K_d \qquad\qquad (A.I.5)$$

The units of f are volume of particles/volume of sea water and, thus, the units of K_d are volume water/volume particles.

Clearly for this concept of scavenging to hold in the oceans, the particles must be in contact with the same parcel of water for a long enough time to reach a sorption equilibrium. Application of the model in Figure A.I–4 then requires knowledge of the k's as well as of particle fluxes and sinking rates, and rates of transformation between small and large particles. Approaches toward determining these parameters and a characterization of existing knowledge are set forth in the following Sections.

3.2.   Determination of scavenging rate constants

Considerable support for the reversible scavenging model described above is provided by the depth distribution of naturally occurring thorium isotopes in the oceans. ^{230}Th (half-life = 75 200 a), ^{234}Th (half-life = 24 d) and ^{228}Th (half-life = 1.9 a) are all produced in sea water by decay of dissolved parent nuclides (isotopes of U for ^{234}Th and ^{230}Th and ^{228}Ra for ^{228}Th). Recent determinations of the Th isotopes in sea water by Nozaki et al. (1981) and Bacon and Anderson

---

[2]   Strictly, Equation (A.I.5) should be written as $\dfrac{k_1}{k_2} = \dfrac{f\,K_d}{(1-f)}$.

In the water column where the fractional volume of particles f $\ll$ 1 this distinction is unimportant but in the sediments the more complete form may be needed (for example, see Appendices V and IX).

TABLE A.I-II

K_d FOR SOME REACTIVE ELEMENTS

IN AMBIENT OCEAN WATER (C_p = 10μg kg^{-1})

Element	K_d [a]
Th	2 x 10^7
Pa	1 x 10^7
Mn	1.0 x 10^7
Pb	8 x 10^6
Pu	5.0 x 10^6
Cu	1.5 x 10^6

[a] calculated assuming a particle concentration of 10 μg kg^{-1}.

(1982) reveal increases in both the particulate and soluble ^{230}Th with depth (see e.g. Figure A.I-5). Such a distribution for the long-lived ^{230}Th would be expected if sinking particles reach reversible sorption equilibrium between dissolved and particulate ^{230}Th. From the partitioning of ^{230}Th and ^{234}Th between particles and solution, two equations containing the rate constants k_1 and k_2 can be written (assumed the same for both isotopes) and the values of k_1 and k_2 can be determined. For the western North Pacific, Nozaki et al. (1981) calculate k_1 = 1.5 a^{-1} and k_2 = 6.3 a^{-1}.

In the eastern tropical Pacific, Bacon and Anderson (1982) determine values for k_1 of 0.2 to 1.3 a^{-1} and 1.3 to 6.3 a^{-1} for k_2. The Bacon and Anderson (1982) data also show a correlation between k_1 and particle concentration as discussed earlier. The desorption rate coefficient k_2 shows less correlation to particle concentration.

The value of K_d which results from Bacon and Anderson's (1982) data is ~ 2 x 10^7. Table A.I-II summarizes K_d values for other reactive elements in the mid-depth open ocean, based on the observed ratios of C_p/C_d and a mid-depth particulate concentration of 10 μg/kg.

The treatment of the thorium isotope data by Nozaki et al. (1981) and Bacon and Anderson (1982) assumes a reversible uptake of thorium on slowly sinking small particles. However, Bacon and Anderson (1982) also explicitly considered the possibility of Th uptake onto particles which are then rapidly removed by rapidly settling larger particles. Their data suggest that the rate constant for Th removal by the large-particle flux is $\lesssim$ 0.1 a^{-1}, substantially smaller than for desorption.

As the preceding discussion has demonstrated, marine geochemists are only now gathering the kind of data necessary to evaluate the kinetics of the interaction of dissolved nuclides with fine particles. That these particles are important in removing reactive elements from the mid-depth open ocean is demonstrated by calculated settling rates of 3 x 10^{-4} to 2 x 10^{-3} cm s^{-1} for particles removing Th and Pb from solution. These rates correspond to particle sizes of 2-5 μm. That the flux of large (i.e. biogenic) particles is also important (at least in open ocean surface waters) is evident from the rather rapid removal and transport of

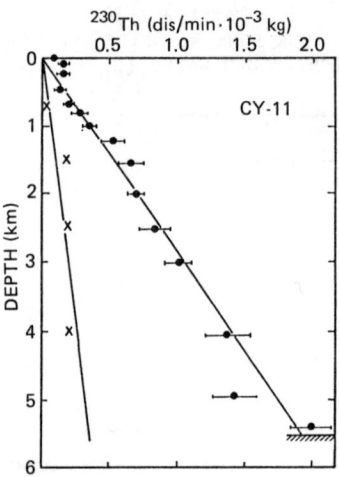

FIG. A.I–5. The vertical distributions of total ^{230}Th (filled circles) and particulate ^{230}Th (crosses) in the North Pacific (after Nozaki et al., 1981).

reactive fallout nuclides (e.g. Pu and fission product isotopes of Zn, Nb, and Ce) from surface waters to the bottom (e.g. Osterberg et al., 1963). The interaction of large particles with dissolved reactive nuclides seems maximum in the surface waters. As these particles sink, a fraction of them may decompose or disintegrate, releasing some scavenged nuclides back to solution. This process may be observed in vertical profiles of nuclides (e.g. ^{210}Po, ^{228}Th, Pu) which have a source or strong production term in surface waters. The Pu distribution in the North Pacific appears to demonstrate these effects quite well. As Figure A.I–6 shows, the N. Pacific $^{239,240}Pu$ profile determined in 1973, about a decade after the major input from weapons testing, has low concentrations in surface waters, a distinct maximum at ≈ 500 m and low mid-depth concentrations with somewhat higher concentrations near the bottom. Sediment inventories of Pu in the area represent 10% of that in the water column. The N. Pacific water column inventories of Pu are themselves higher than the latitudinal global fallout averages due to close-in fallout from the Pacific tests. The fraction of Pu on filterable fine particles is generally less than 10% and the filterable Pu profile resembles the dissolved Pu profile. Sediment trap samples which, unlike filtered samples, catch the large particle flux, show increasing Pu concentrations to the bottom. Thus the effect of the flux of particles on the development of the Pu transient can be summarized as (1) relatively rapid incorporation into large particles formed in surface waters, (2) subsurface releases of Pu from the sinking large particles, (3) transport of a relatively small fraction of Pu to deep water with a partial regeneration at or near the sediment/water interface, and (4) uptake of Pu remaining in the water column onto the slowly settling fine particles or possibly onto large particles.

A more rigorous application of the model of Figure A.I–4 requires information on the rate of destruction or decomposition of large

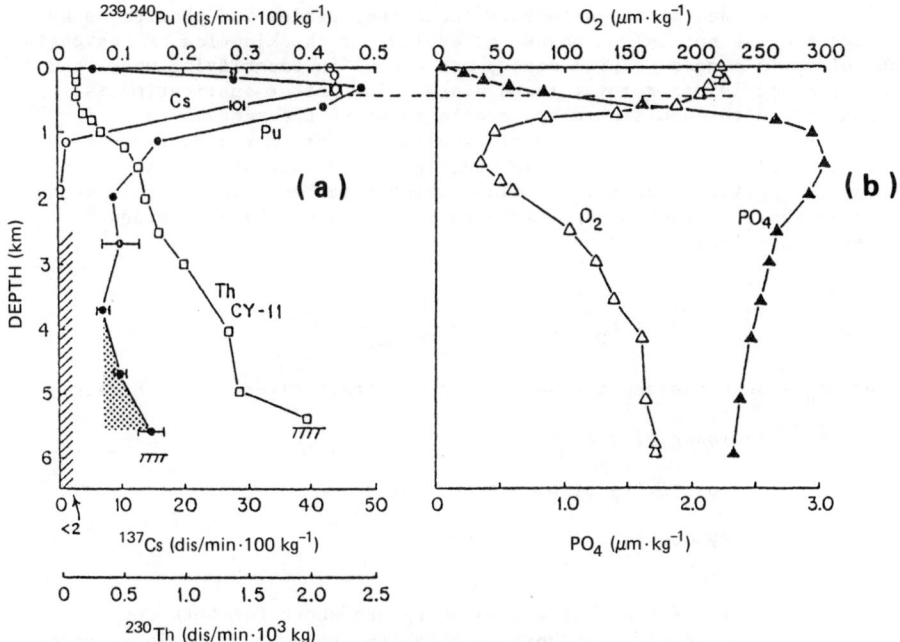

FIG. A.I-6. (after Sholkovitz, 1983).
(a) ^{239,240}Pu, ^{137}Cs and ^{230}Th concentrations versus depth at GEOSECS station 225 in the North Pacific. Plutonium and caesium data from Bowen et al. (1980), ^{230}Th from Nozaki et al. (1981), (b) O_2 and PO_4 concentrations versus depth at GEOSECS 225.

particles, the rate at which settling large particles can impact or sweep up small particles, the rate constants for adsorption and desorption onto large and small particles and the sinking rates for small and large particles. The sinking rates and sorption rate constants for the small particle population are inferred from natural radionuclide studies. Uptake and release rates onto large particles have so far been determined principally from laboratory studies. For example, Fowler (1982) tabulates release rates for Zn, Hg, Se, Pu, Cs, and Po from euphausid faecal pellets, molts, and carcasses. Half-times for release are generally < 20 days, with faecal pellets somewhat more retentive on the average.

The rates of disintegration, dissolution, or decomposition of large biogenic particles are relatively poorly known. These rates may be expected to vary with depth, either due to temperature-dependent microbial reactions (i.e. organic matter oxidation) or to the rate of dissolution of tests.

3.3.   Scavenging regimes

Nuclides of the naturally occurring uranium and thorium decay series provide not only a means for evaluating the kinetics of scavenging but also, in a more general sense, are useful in identifying where and on what time-scale net removal takes place. The latter application is accomplished through simple box models in which the removal is parametrized by mean residence times with respect to radioactive decay and non-decay processes (i.e. scavenging and deposition). Table A.I-III shows the particle-reactive nuclides which have been used in this sense. Because their production and decay rates are known, "mean residence times" are calculated from

$$ t_D = \frac{A_D}{\lambda_D (A_P - A_D)} $$

where t_D = mean residence time of daughter (particle-reactive nuclide)

A_D = radioactivity of daughter nuclide

A_P = radioactivity of parent nuclide

λ_D = decay constant of daughter nuclide.

Table A.I-III shows that values of t_D are short for both the surface open ocean and nearshore relative to the deep sea. This approach permits the conclusion that net removal is most rapid in the surface and nearshore waters, but does not, of course, provide information about possible biological or chemical recycling of the nuclides involved.

The importance of different oceanic regimes (e.g. continental margins, deep-sea sediments) in accounting for net accumulation of particle-reactive nuclides can also be addressed through the uranium and thorium decay series nuclides. This is accomplished by comparison of net removal from the overlying water column (i.e. the total deficiency of the daughter nuclide relative to that expected for radioactive equilibrium with its parent) and the integrated amount of daughter nuclide in the underlying sediments. If the flux or inventory of the nuclide matches that expected from its deficiency in the water column, a case can be made for scavenging operating in a predominantly vertical sense. Deviations from such a vertical balance can often be attributed to horizontal processes, acting either in the water column or on the bottom sediments.

Measurements of both [230]Th (half-life = 75 200 a) and [231]Pa (half-life = 32 000 a) in sea water show that $\geq 99\%$ of the production of these nuclides from U parents is removed from the water column by scavenging. However, deep-sea sediments commonly have less than 70% of the expected inventory of [230]Th (Kadko, 1981). The sediment trap data of Anderson (1981) show that the fluxes of [230]Th in open ocean Pacific and Atlantic sites are less than predicted from U decay in the water column. The deviation is even more striking for [231]Pa. The pattern suggests that Pa has a longer mean residence time in the water column and is incompletely removed by vertical scavenging. The implication is that Pa can be transported in association with the oceanic circulation to sites of more intense scavenging. Some of these areas have been identified. In general, they are associated with a high particle flux

TABLE A.I-III

MEAN RESIDENCE TIMES (WITH RESPECT TO NET REMOVAL FROM PROCESSES
OTHER THAN RADIOACTIVE DECAY) FOR PARTICLE-REACTIVE NUCLIDES
OF THE U AND Th DECAY SERIES

Soluble Parent	Particle-reactive daughter	Daughter half-life	Open Ocean		Nearshore
			Surface (a)	Deep (a)	(a)
^{238}U	^{234}Th	24 d	0.4[1]	>>0.25[2]	< 0.3[10]
^{228}Ra	^{228}Th	1.9 a	0.5[3] 0.5-1 (Eq. N. Atlantic)[11]	--	< 0.3[10]
^{234}U	^{230}Th	75 200 a	--	17-41[4]	--
^{235}U	^{231}Pa	32 000 a	--	34-130[4]	--
^{226}Ra	^{210}Pb	22.3 a	1.4 - 2.3 (N. Atlantic)[5] 1.7 (N. Pacific)[6] 0.5 - 1.0 (NW Pacific)[7]	~50 (Pacific)[8] ~100 (S Pacific)[9] < 96 (NW Pacific)[7]	< 0.3[10]
^{210}Pb	^{210}Po	138 d	0.3 - 0.6 (N Atlantic)[5] 0.6 (N Pacific)[6,7]	4 (N Atlantic)[5] 2 (S Pacific)[9]	

References
(1)   Matsumoto (1975)
(2)   Amin et al. (1974)
(3)   Broecker et al. (1973)
(4)   Anderson (1981)
(5)   Bacon et al. (1976)
(6)   Nozaki et al. (1976)
(7)   Nozaki and Tsunogai (1976)
(8)   Craig et al. (1973)
(9)   Thomson and Turekian (1976)
(10) Li et al. (1981)
(11) Li et al. (1980)

either through intense biological productivity or input of detrital
sediment. DeMaster (1979) has identified the rapidly accumulating
sediments of the South Atlantic/Antarctic as a site of preferential Pa
removal and Anderson (1981) has observed substantially greater than
expected Pa fluxes in the Panama Basin. A pattern of enhanced scavenging
in continental margins has emerged from other studies as well. Data
supporting this pattern include:

(1)   decreasing scavenging mean residence times of ^{234}Th (Bhat
      et al., 1969), ^{228}Th (Broecker et al., 1973) and ^{210}Pb
      (Bacon et al., 1976) as a coast or oceanic boundary is
      approached (see Table A.I-III).

(2)   within a coastal environment like the New York Bight, Th
      isotopes show decreasing mean residence times as the particle
      concentration increases (Li et al., 1979).

(3)   Substantially higher than predicted inventories of ^{210}Pb,
      ^{228}Th and Pu are observed in sediments from upwelling areas
      on oceanic eastern boundaries (see Table A.I-IV and
      discussion below).

ACCUMULATION OF REACTIVE NUCLIDES IN SELECTED AREAS
OFF THE WEST COAST OF NORTH AMERICA

	Nuclide	Predicted Flux (dis/min cm^{-2} a^{-1})	Measured Flux (dis/min cm^{-2} a^{-1})	Reference
Santa Barbara Basin	^{234}Th	1.1×10^6	1.4×10^6	(1)
	^{228}Th	4×10^2	1.5×10^3	(1)
	^{210}Pb	7.5×10^3	5.0×10^4	(1)
	^{239,240}Pu inventory (mCi km^{-2})	2.0	4.5	(2)
Washington Shelf	^{210}Pb	1.5	4.9	(3)

References: (1) Moore et al. (1981)
(2) Sholkovitz (1983)
(3) Carpenter et al. (1981)

The texture of oceanic scavenging which these studies illuminate implies that horizontal processes are important in redistributing particle-reactive nuclides from mid-ocean areas characterized by slower scavenging rates to areas such as the continental margins where the rates of scavenging and removal are faster. Spencer et al. (1981) have attempted to model the ^{210}Pb distribution in the North Atlantic on the basis of a flux of ^{210}Pb to the ocean boundaries (see Figure A.I-7). With such a model, ~ 16% of the non-radioactive decay removal of ^{210}Pb took place at the ocean margins. Spencer et al. (1981) postulated that one possible explanation for higher scavenging rates at ocean margins is the reduction of Fe and Mn in association with organic matter oxidation in margin sediments. The flux of reduced Fe^{2+} and Mn^{2+} to the overlying water, followed by oxidation to Fe^{3+} and Mn^{4+} and precipitation as oxides, can serve as an efficient scavenging mechanism for particle-reactive nuclides such as ^{210}Pb. Spencer et al. (1981) also noted that the North Atlantic ^{210}Pb distribution can be fitted equally well by a model which allowed particle concentrations to vary by a factor of about 40 from the ocean interior to the boundary (specifically, the western boundary). Such increases are not observed in the North Atlantic, and a model incorporating a reactive boundary and a less reactive oceanic interior may be the more realistic one.

An excellent example of enhanced accumulation of reactive nuclides on continental margins is seen off the west coast of North America. Inputs of natural series nuclides and Pu to sediments of the Santa Barbara Basin and the Washington shelf are several times that expected from supply from either the atmosphere (^{210}Pb, Pu) or in situ production (^{210}Pb, ^{228}Th). These results are given in Table A.I-IV. Only in the case of the short-lived isotope ^{234}Th does the measured flux approximately equal the predicted flux. The mechanism by which these excess

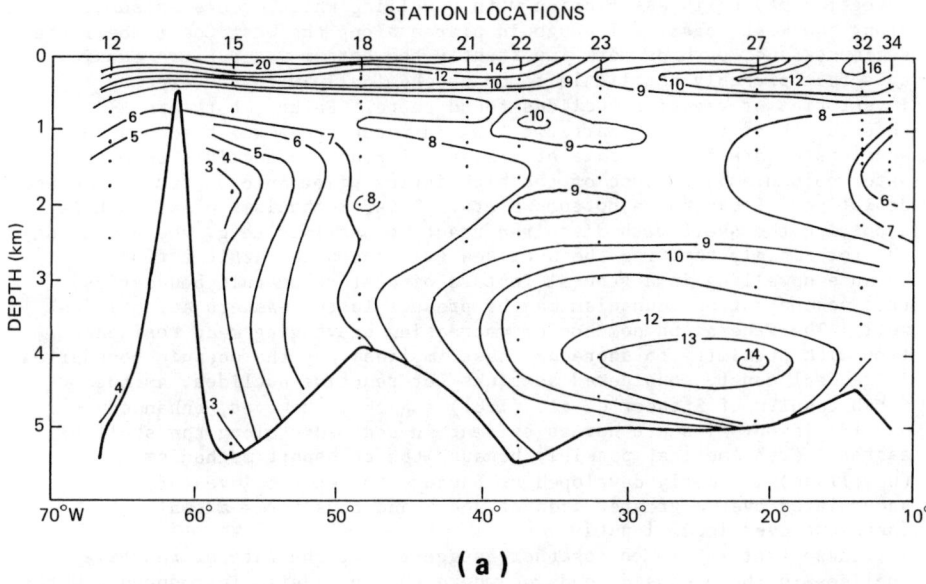

(a)

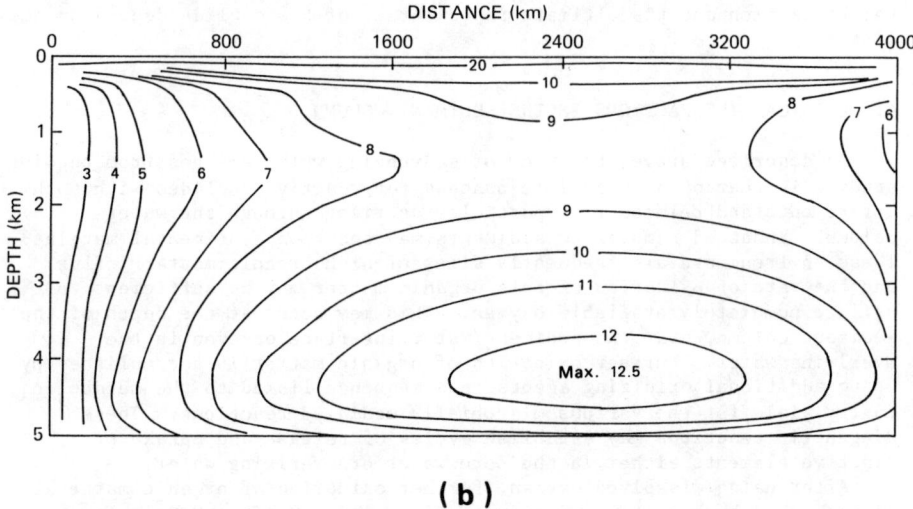

(b)

FIG. A.I–7. (from Spencer et al., 1981).
(a) Dissolved ^{230}Pb (dis/min·100 kg^{-1}) in west-east section at about 15°–30°N in the Atlantic Ocean,
(b) Model ^{230}Pb (dis/min·100 kg^{-1}) distribution resulting from a numerical solution of Equation (1) with ^{230}Pb scavenging at the side and bottom boundaries.

inventories are generated is probably advection of water from offshore which therefore can have higher concentrations of the nuclides. The advection may be in association with upwelling which occurs seasonally along the West Coast. Although in places along the West Coast Shelf the sediments and, perhaps more importantly the bottom waters, are anoxic, this fact probably has little to do geochemically with the excess inventories of reactive nuclides found there. Rather, both are a consequence of the high particle flux in these areas which serves to accelerate markedly the rate of removal of reactive nuclides from the water column and, because of the high influx of organic matter, may cause depletions of oxygen in bottom waters. Thus, mechanisms exist both for supplying the shelf with dissolved reactive nuclides (e.g. via upwelling) and for rapidly removing the nuclides (via the high particle flux). Because upwelling is a general feature of eastern oceanic boundaries, this concentration mechanism may be present in the eastern Atlantic as well. The general phenomenon of decreasing scavenging mean residence time with proximity to shore or coast implies that the oceanic boundaries in general can be considered as sinks for reactive nuclides, as the ^{210}Pb results of Spencer et al. (1981) suggest. However, enhanced nuclide inventories are not as evident in sediments along the shelf of eastern North America, possibly because the transport mechanism (upwelling) is poorly developed or because the excess inventory is accumulated over a greater area of shelf and thus forms a smaller increment over local inputs.

These factors, taken together, suggest that the fate of reactive nuclides in the oceans depends on where the nuclide is introduced and the fate of the particles with which the nuclide interacts. The relative rate of uptake onto particles depends on the former while possible remobilization and the ultimate repositories of the nuclide depend on the latter.

4. CHEMICAL INTERACTIONS IN THE SEDIMENT COLUMN

As described above, the view of scavenging which emerges from ongoing studies is that of a reversible process for reactive nuclides with both uptake onto and release from particles settling through the water column. Enhanced removal to sediments may occur on continental margins. These environments are frequently sites of high organic matter influx, and the rate of oxidation of this organic matter may be sufficient to utilize completely available oxygen. This may occur at the depth of the sediment column, near the sediment/water interface or even in the overlying water. Further oxidation of organic matter is accomplished by using additional oxidizing agents in a sequence linked to the metabolic energy yield for the various microbially mediated reactions. These diagenetic reactions may establish cycles of release and uptake of reactive elements either in the pore water or overlying water.

After using dissolved oxygen, further oxidation of organic matter is accomplished by reduction of nitrate, manganese (Mn^{4+} as MnO_2 to Mn^{2+}), iron (Fe^{3+} as $Fe(OH)_3$ to Fe^{2+}), sulphate and finally methane formation by disproportionation of organic matter. This sequence of reactions is graphically displayed in many nearshore marine sediments and is seen as changes in the pore-water concentrations of the various oxidizing agents. Nitrate and sulphate become depleted in the pore water. Reduction of Mn^{4+} and Fe^{3+}, both of which are relatively insoluble as oxy-hydroxides, leads to accumulations of the more soluble

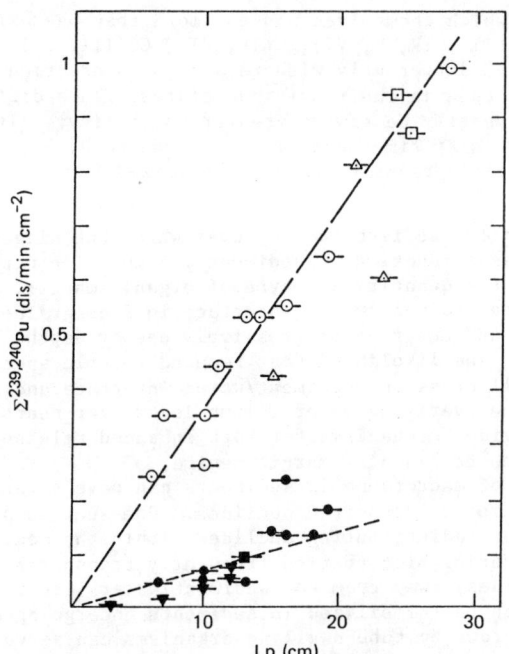

FIG. A.I–8. $^{239,240}Pu$ *inventories in coastal marine sediments versus the depth (Lp) where* $^{239,240}Pu$ *activities are below 5% of the surface activities (Santschi et al., 1980).*
Nearshore Cores (h < 100 m):
○ *Buzzards Bay, Gayhead (Livingston and Bowen, 1979)*
△ *Narragansett Bay*
□ *N.Y. Bight*
Offshore Cores (h > 100 m):
● *Wilkinson Basin and Georges Bank (Livingston and Bowen, 1979)*
■ *N.Y. Bight*
▼ *Atlantic Cores (Noshkin and Bowen, 1973).*

reduced Mn^{2+} and Fe^{2+} in the pore water. Sulphate reduction results in increasing concentrations of sulphide in the pore waters, and this sulphide effectively precipitates reduced iron as iron monosulphides and ultimately pyrite (FeS_2). Many other metal sulphides are similarly insoluble. Reduced manganese may precipitate as $MnCO_3$ as alkalinity increases.

The redox gradient established by these diagenetic reactions has several consequences for the fate of particle-reactive contaminants.

(1)   Nuclides may be released to solution as the Mn and Fe oxides to which they are sorbed are solubilized during reduction. Additional precipitation reaction such as of metal sulphide may re-scavenge these nuclides.
(2)   Depending on their susceptibility to bacterial oxidation, organic contaminants may be appreciably altered.

(3)   Species which themselves possess more than one oxidation state
        (e.g. Pu(III, IV, V, VI), U(IV, VI), Cr(III, VI) may be reduced
        and behave differently with regard to interaction with particle
        surfaces than in their oxidized states. This difference will
        not necessarily be toward greater solubility. Although the
        evidence is at times conflicting, reduced U, Pu, and Cr may be
        more particle reactive than the oxidized forms.

    The depths in the sediment column over which the diagenetic reactions
take place vary as a function of sediment accumulation rate and, in
particular, with the quantity and type of organic matter. The reactions
develop nearer the sediment/water interface in fine grained continental
margin sediments and occur at progressively deeper depths in the sediment
column offshore. The likelihood that reduced soluble species such as
Fe^{2+} and Mn^{2+} will cross the sediment/water interface and be
re-oxidized in the overlying water column is greater nearshore. This
process thus provides mechanisms for both enhanced release and scavenging
of nuclides in the continental margin regime.
    The presence of macrofauna in sediments can have several consequences
for the distribution of deposited nuclides. One such consequence is that
of transporting or redistributing nuclides within the sediment column.
Particle mixing during bioturbation frequently transports particles
(+ adsorbed nuclides) away from the sediment/water interface. For
nuclides which can be remobilized in sediments undergoing diagenesis, the
irrigation of burrows by tube-dwelling organisms can serve as a mode of
transport out of the sediment. Indeed, the presence of burrows
penetrating the sediment possibly to depths of decimetres can drastically
shorten the effective path length over which a chemical species must
diffuse to reach an interface. Depending on burrow density, this radial
diffusion geometry around burrows can radically alter pore-water profiles
of species such as Fe^{2+}, Mn^{2+}, NH_3, and HCO_3^- and enhance the
fluxes to the overlying water.
    For strongly particle-reactive nuclides, particle mixing by infauna
can actually serve to enhance the inventory of the nuclide in the
sediment column. Figure A.I-8 shows one illustration of this concept.
The $^{230,240}Pu$ inventory in nearshore sediments is well correlated with
the depth to which the Pu is distributed (which is a measure of the depth
of biological mixing, Santschi et al., 1980). The pattern appears to be
developed, although less clearly, in cores from deeper water as well. A
similar pattern has been established for ^{210}Pb in a single depositional
basin (Long Island Sound, Turekian et al., 1980). Thus if other factors
(such as grain size) are approximately equal, a greater intensity and
depth of particle mixing results in enhanced transfer of nuclides to
depth in a deposit. Because there is no net flux of particles, this
implies that particles with high concentrations of a reactive nuclide
will be mixed downward and exchanged for particles with low nuclide
concentrations mixed upward toward the sediment/water interface.
Resuspension of interface particles by physical or biological means
allows these particles to either scavenge more nuclide from the overlying
water column or to mix with particles with high nuclide concentrations
from adjacent areas where bioturbation may be less intense. The result
is an enhanced flux of nuclide (but no net flux of particles) into the
areas of intense biological mixing.
    Thus there may be both chemical and physical (i.e. bioturbation)
redistribution of a nuclide in the sediment column. Predicting the fate

70

of such a nuclide after deposition requires information on its chemical
behaviour in the light of the diagenetic reactions occurring in the
sediment and adequate characterization of the nature and rate of
biological mixing in the deposit. The depth distribution of short-lived
particle-reactive nuclides in the sediment column can provide a
quantitative measure of the rate of biological mixing of the upper
decimeter of the deposit. Values generally decrease toward the deep sea,
reflecting the decreasing rate of mixing (probably a result of generally
decreased metabolic activity) in the deep-sea environment.

Appendix II

NATURAL HISTORY OF THE OCEAN

The spatial distribution of organisms in the world ocean tends to be
ordered into a recognizable broad pattern of zonation (Figure A.II-1 and
Table A.II-I). This is relevant to deep-ocean dumping strategies because
identifiable rate processes involved in the biological transfer of
material through the ecosystem are associated with such zonation.
Sufficient knowledge of some of the rates of these processes exists to
allow their use in models of the fate and transfer of contaminants
disposed of in the deep sea.

The most extensive region of the world ocean is the mid-oceanic
central gyre environment over abyssal and hadal depths (about 85% of the
world ocean area), bounded around its rim by the continental slopes
rising to the shallow continental shelf seas. Vertically the oceanic
environment is divisible into epipelagic (0-150 m), mesopelagic
(150-1000 m), bathypelagic (1000-2500 m), abyssopelagic (2500 to
about 100 m above the sea floor) and benthopelagic (within about 100 m of
the sea floor) zones and, bounded by the land/sea interface, the benthic
zone.

1. EPIPELAGIC (= PHOTIC) ZONE

This zone is bounded above by the air/water interface with its
specialized fauna in the upper few centimetres. The lower boundary is
somewhat arbitrarily chosen to be 150 m depth, the maximum depth at which
the intensity of sunlight is sufficient for the photosynthetic process of
the phytoplanktonic producers. (This level also conveniently
approximates to the mean depth of the outer edge of the Continental
Shelf.) Between these limits the phytoplankton synthesizes the organic
matter that becomes the prime energy source of the marine environment.
Generally the warm central waters are low both in nutrients and standing
stocks of phytoplankton. Similarly stable regions support communities
with low biomass, but with high species and trophic diversity (with
carnivores dominating), whereas communities in regions of intense
divergence have a higher biomass with low species and trophic diversity
(with coarse filter-feeding organisms dominating). It is estimated that
the production of the open ocean is about 15×10^9 t carbon/year. Of
this about 75% is utilized in the upper 300 m while 20% passes through to
meso- and bathypelagic levels and to the bottom. About 5% reaches
bathypelagic levels and deeper, with an almost equal division between
pelagic and benthic consumption, although the latter is concentrated
vertically on the bottom (Riley, 1970). The overall transfer of
production from one trophic level to the next has been shown to be about
10% for primary consumers and about 15% for subsequent stages (Ricker,
1969). Thus the more links in the food chain the less the terminal
productivity. It has been suggested that typically there are about five
links in the food chain, but some chains are known to be shorter (e.g.
phytoplankton - krill (Euphausia superba) - fin whales) and thus probably
more efficient.

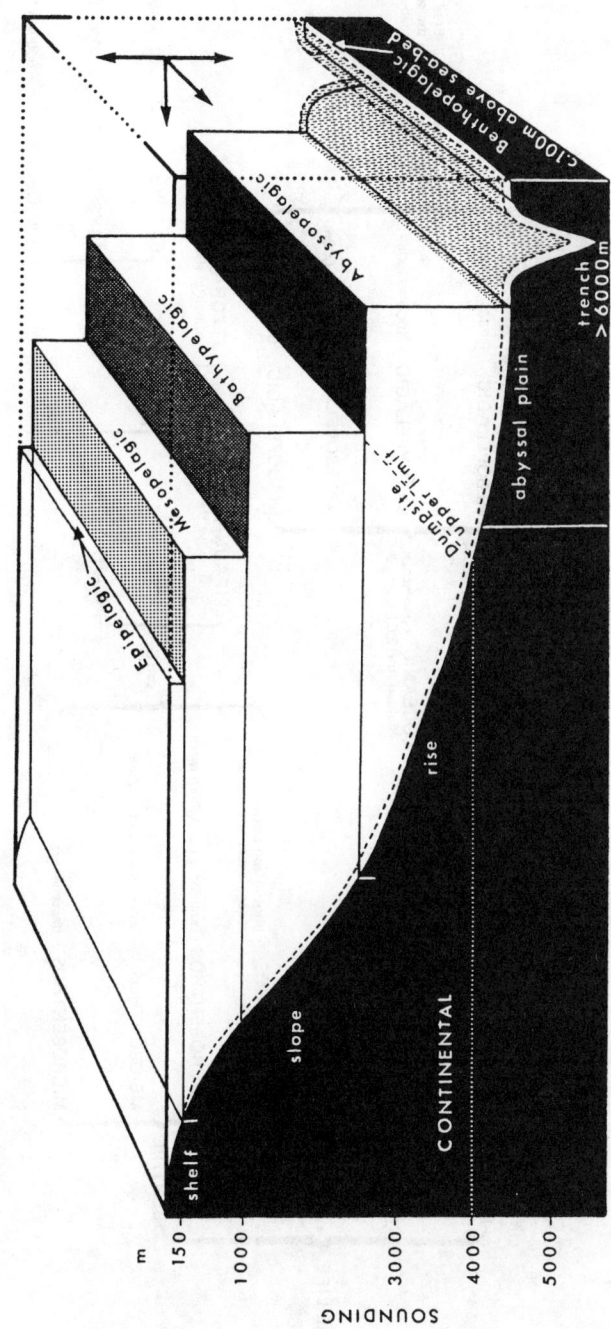

FIG. A.II–1. *Three-dimensional representation of vertical zonation in the deep sea. N.B. The different zones cannot be rigidly separated one from the other. (Potential multi-directional pathways of biological flux are indicated by the arrows at the right of the diagram.)*

73

TABLE A.II-1. A DESCRIPTIVE CLASSIFICATION OF THE MARINE ENVIRONMENT.
(Depth ranges must be regarded as approximate and the zones cannot be rigidly separated: cf. Figure A.II-1.)

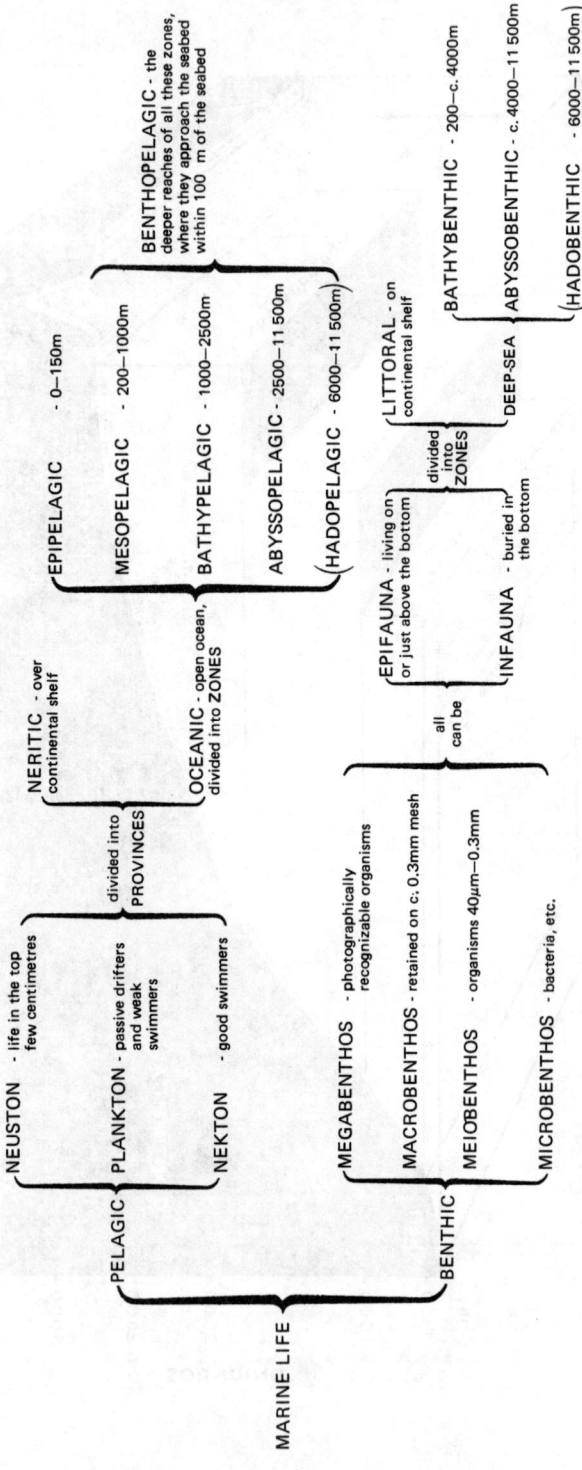

Non-migrant and nocturnally vertically migrating grazers (planktonic and nektonic) depend directly on the phytoplankton. Non-migrant larval forms, largely from adult stocks at lower levels, contribute conspicuously to the planktonic biomass. Notable among these are fish larvae from all levels in the water column to benthopelagic and benthic species from abyssal depths. As they develop, they undergo a downward migration to the upper levels of the adult living space (see below). Carnivores, both non-migrant and migrant, of all sizes to the largest whales prey on the grazers, and many on one another. All contribute faecal waste which is largely utilized by detritivores lower in the water column. Recent work has shown a considerable variation in the sinking rates of faecal pellets, from 5-220 m d^{-1} from copepods, 43-862 m d^{-1} from euphausiids, 440-1800 m d^{-1} from pteropods, 41-208 m d^{-1} from doliolids and 320-2700 m d^{-1} for those from salps (see Angel, 1983) for review. The high sinking rates from pteropods and salps suggest that they may be disproportionately important in the flux of biogenic material during periods when they form dense population blooms. Contributing also to this flux are the carcasses of all classes of organisms which, as a result of natural mortality, form a food source for deeper living communities. Those from the largest (e.g. tuna, marlin, sharks and cetaceans) are the 'large lumps' which may descend as isolated food packages, of considerable size, to the deep-sea floor.

Reverse migrations, both diurnal and intermittent, are carried out by occupants of this zone. Notable among the latter are whales. Baleen whales may dive some tens of metres to feed on krill, while the toothed whales (e.g. sperm whales) dive to greater depths (possibly even to about 3000 m, but usually to less than 400 m for periods of up to 10 minutes with only a few dives (c.5%) exceeding 800 m and 30 minutes duration to feed (benthopelagically ?) on cephalopods. Based on the unexploited (1946) stock of sperm whales, the total population of 2 x 10^6 probably consume 260 x 10^6 tons of cephalopods annually (Clarke, 1977). (Independent estimates of annual potential catch of oceanic cephalopods project it to be as high as 500 x 10^6 tonnes, emphasizing the importance of this exceptionally poorly sampled group.) Large fishes (e.g. sharks, swordfish, billfish and tuna) carry out (irregular) vertical excursions between this zone and the mesopelagic layers below. These fishes comprise the largest oceanic fishery, mainly in tropical and sub-tropical waters, with a current yield of about 3 x 10^6 t (Yearbook, FAO 1979).

2. MESOPELAGIC ZONE

Concentrations of both dissolved and particulate organic matter vary widely in the upper waters but at depths below 200-500 m there is relatively little change over most of the ocean. Thus such uniformity is characteristic of the mesopelagic zone, which extends from about 150 m down to the limit of the attenuation of solar light in the clearest oceanic waters - at around 1000 m - and incorporates the waters of the permanent oceanic thermocline. This is the zone of the vertically migrating grazers and carnivores, stacked in predictable distribution layers, with a standing crop estimated at around 6 g carbon m^{-2}. Migrants feed generally at night, often in the epipelagic or shallow mesopelagic zones, and many occupy complicated niche patterns during both daylight and darkness. It has been estimated that there is a vertical

movement of around 25 tonnes $km^{-2}d^{-1}$ among organisms residing in the
upper 350 m of the water column (Longhurst, 1976). It has been suggested
that, if epipelagic predators such as tuna do not feed at night, they
cannot retrieve the energy won by the migrating mesopelagic fauna. This
may be seen as an 'energy valve' permitting the downward transfer of
energy but reducing the upward one. If such a mechanism exists in other
parts of the food chain, then possibly it is fundamental to the support
of a relatively large biomass in the deeper layers. Generally the size
of the dominant organisms in this zone is small (<10 cm in length), which
is consistent with estimates that, in energetic terms, midwater predators
must not be much larger than their prey. Another important element of
the mesopelagic fauna is the non-migrant detritivores depending on the
faecal rain from above, which themselves fall prey to migrant and
non-migrant predators alike.

3. BATHYPELAGIC ZONE

     Situated below the permanent thermocline and the greatest extent of
solar illumination, this zone may be considered the largest environment
on earth, changeless in its physical surroundings. The vertical
distribution ranges of few diurnal migrants extend into this zone and the
plankton is probably largely coprophagous, detrivorous or carnivorous.
On the other hand, the micronekton consists predominantly of carnivores
or detrivores, often highly adapted to a sparse food supply. For
instance in general terms, respiration rates per unit body weight of
species from 1000 m are about 1/10 those of shallow water forms. Indeed,
among bathypelagic fish there is a reduction of structural complexity in
response to the selective pressures of this environment (Marshall,1979).
Estimates of the standing crop confirm the depleted biomass of the
bathypelagic zone, mirrored in the low faunal abundance and diversity.
That at 1000-2000 m is around 1.5 g carbon m^{-2}, while from 2000-4000 m
it is estimated to be 0.5 g carbon m^{-2}. Recent results of deep pelagic
sampling indicate that the reduction of biomass with depth can be
considered to be exponential, although an abrupt decrease in fish biomass
occurs around the 2500 m level and extends down to some tens of metres
from the bottom (Angel and Baker, 1982), to suggest a rationale for
distinguishing an even more impoverished subdivision, an abyssopelagic
zone. Such a marked depletion in biomass below 2500 m, together with the
evident absence of diurnal migrants below 1000 m, indicates the degree of
isolation of the near-surface fauna from abyssal benthic influence.
Indeed, conversely, it seems that one of the few predictable contacts
bathypelagic animals have with the more productive waters above their
habitat is through migration of eggs and larvae during their early life
history. Certain weak vertical zonation patterns are evident in the
bathypelagic zone. For instance, only two species of ostracod out of
about 45 span the total depth range from 1500-3900 m in the eastern North
Atlantic.

4. BENTHOPELAGIC ZONE

     The living space of benthopelagic animals is near the deep-sea floor
and so, near the walls of the abyssal basins, such fauna may well merge
with those of mesopelagic origin over the upper continental slope at

depths of 200-1000 m and with the most abundant species of bathypelagic origin at depths greater than 1000 m. Fish, especially rattails, are prominent members of this fauna, both in abundance and species diversity. Furthermore, there is evidently no such loss of structural complexity among benthopelagic fishes as is found in their bathypelagic counterparts. Benthopelagic fishes are most abundant in slope areas. The oceanic rim between 200-2000 m soundings provides a little less than 9% of the area of the ocean, yet it may well bear three-quarters of the entire benthopelagic and benthic fish fauna. For instance, in the eastern North Atlantic, fish abundance in 200-2000 m soundings has been found to be 100-10 per 1000 m^2 and in 2000-5000 m, 1.0-0.1 per 1000 m^2. Similarly, over the same sounding ranges, fish biomass is 10-1.0 g m^{-2} and 1.0-0.1 g m^{-2} respectively (Institute of Oceanographic Science, unpublished). In species richness, too, the slope regions are prominent. Of 40 species of rattails in the western North Atlantic 24 have centres of abundance in 200-1000 m soundings, 11 at 1000-2000 m and 5 at depths greater than 2000 m. Akin to their pelagic relatives, benthopelagic fishes, in general, display discrete limits of vertical distribution relative to sounding, although limited evidence exists to indicate considerable excursions may be made into the overlying waters. Typically such ranges are of the order of tens to a few hundreds of metres on the slope but they may extend to > 1000 m, mainly in species living on the continental rise and abyssal floor. Generally, within the living depth of any particular species, the smaller forms (juveniles and adolescents) dwell in the shallow parts of the distribution range, while the adults tend to occupy the lower reaches. The consistency of these observations in sampling suggests that probably the adult range of movement is less than the overall living depth of the species.

There is no evidence, as yet, to suggest that there are conspicuous up-slope migrations among benthopelagic fishes as occur in some squid. Todarodes sagittatus annually ascends the slope into shallow shelf regions, thus introducing nutrients at a high trophic level into shallow water (Clark, 1977). Little is known about the diet of such species of squid. Nevertheless, the distribution of the catches of sperm whales which prey on them suggests that they may well be most densely distributed in areas of upwelling (Cushing, 1971). Evidence from gut contents of slope-dwelling benthopelagic fish species indicates that the majority depend completely or in part on pelagic food consisting of the meso- and bathypelagic fauna impinging on the shoaling sea floor. Thus it is not surprising that the peak biomass of pelagic organisms is matched among benthopelagic fishes, occurring around the mid- to upper slope (see above) (Marshall and Merrett, 1977). Dominant abyssal benthopelagic fishes whose diet is known (e.g. Coryphaenoides (Nematonurus) armatus) also depend on pelagic (or benthopelagic) organisms. (The smaller C. (Coryphaenoids) guentheri (vertical distribution 1100-2800 m) on the other hand, processes the substrate for food and, incidentally, has been found to contain 3-4 times more of the naturally occurring radionuclide pollonium (^{210}Po) in its tissues than C. (N.) armatus.)

Very little is known about the reproductive strategies and early life-histories of deep-sea bottom-living fishes. Much can be inferred, however, from the morphology of both the adults and their eggs. In addition, larvae collected at the surface can, in some instances, be identified with adults of bottom-living fishes. Hence, cartilaginous fishes (sharks, rays and chimaeras) are known to produce young which develop at depth, while many teleosts (i.e. bony fishes) can be safely presumed to have an epipelagic development prior to an ontogenetic migration down to the adult living depths. Notable among the latter are the diverse and abundant rattails (family Macrouridae), whose eggs are known to be buoyant. Similarly the eels, notacanths and halosaurs all have pelagic near-surface leptocephalus larval stages. The ophidioid fishes, on the other hand, have benthic larvae and are often viviparous.

5. BENTHIC ZONE

There are three fundamental regions of the ocean floor; the continental margins, the ocean basin floors and the mid-ocean ridge systems. Over most of the continental slopes there is a covering of terrigenous muds, debris weathered from the land mixed with pelagic ooze. The high content of organic matter contained in these muds is closely related to the nearness to land. The ocean basin floors are less enriched and least of all are the steep-sided mid-ocean ridges with only a thin covering of sediment. Yet communities of animals occur over the entire deep-sea floor. Detrital particles, faecal material and animal carcasses derived from both sea and land sustain deep-sea metabolism. This input may be seasonal in temperate zones where flocculent material is found to accumulate close to the seabed soon after the onset of the spring bloom in the surface waters.

There is a broad size spectrum of organisms in the benthic food web. At the lower end of the scale, bacteria live in the sediments at all levels of the ocean. In soundings of 4000–10 000 m the standing crop of bacteria seems to be about 10^6 g^{-1} sediment. In addition, this microfauna contains yeasts, fungi, amoebae and ciliates. Next in the size range of organisms come the members of the meiofauna (c. 40 μm – 0.3 mm), a diverse group of invertebrates living in the surface sediments at all levels. In numbers and biomass the dominant elements of this fauna are nematodes and foraminifera, at all depths. The macrofauna (>0.3 mm) are next in order of size of benthic organisms. This group consists largely of polychaete worms, peracarid crustaceans and bivalve molluscs, which are mostly permanent or semi-permanent members of the infauna. Their modes of feeding together embrace the predatory, suspension/deposit and filter-feeding habits. The largest invertebrates, the megafauna, are those sampled by large trawls. Many are epifaunal and can be recognized in bottom photographs. This group also includes some octopod molluscs and fish. As with the meiofauna and macrofauna, there are deposit-feeders, suspension-feeders, carnivores and scavengers. Many of the suspension-feeders are attached (e.g. crinoids and sea squirts) and require a reasonable current bearing sufficient food to sustain them. They are the dominant forms on the exposed slopes of sea mounts and mid-ocean ridges. Deposit-feeders (e.g. sea cucumbers, irregular sea urchins, some sea stars, brittle stars, gastropods and various worms) are necessarily limited to soft sediments, where they dominate. Indeed, the benthic biomass is directly proportional to the primary productivity of the overlying surface waters and decreases exponentially with depth. In the mid-gyre regions where surface primary productivity is low, as is the sedimentation rate, oligotrophic conditions prevail to favour the deposit and suspension- feeders, relative to carnivores and scavengers. On the other hand, eutrophic conditions occur under areas of high surface primary productivity, where all benthic feeding types occur in quantity. Once again, however, all are most abundant and diverse on the continental slopes. Benthic fish form part of the group of mobile carnivores (along with molluscs and sea stars, for example). Without the facility of neutral buoyancy, they rest on the bottom and are apparently 'sit and wait' predators. One of the more successful groups, the tripod fishes, have evolved elongated and strengthened rays of their pelvic and caudal fins to set them up somewhat above the bottom, presumably to feeding advantage.

Notable within the overall benthic environment are two specialized habitats. The first is found in the deep-sea trenches (c. 6000-10 000 m soundings) which occur at the shear zones between sinking oceanic plates and overriding plates. Evidently trenches far from land masses in poorly productive areas carry a low standing stock of benthic animals, in contrast to those near land masses and beneath eutrophic waters. The megafauna at least is dominated by deposit feeding organisms but, due to high sedimentation rate and seismic activity, species richness among certain groups (e.g. polychaetes) is low. Nonetheless, though unstable, trenches may well support a high diversity of megafaunal invertebrates, yet generally estimates of benthic biomass are low. The second notable habitat involves energy, not originating from the sun through photosynthesis, but from that derived from the earth along rifts between tectonic plates of oceanic crust. Sea water enters newly formed fissures, where it is heated and chemically enriched with hydrogen sulphide, among other things. It is then ejected upwards through hydrothermal vents. Autochemotrophic bacteria metabolize the sulphide, flourishing to provide a food source for suspension-feeding invertebrates. Such vents sustain a unique fauna of individuals of often exceptionally large size (Hessler, 1981).

6. HORIZONTAL ZONATION AND DISTRIBUTION OF EXISTING FISHERIES

Horizontal discontinuities in the biota of the world ocean are readily apparent, but are less regular than zonation in the vertical plane and generally less trenchant than those found around the continental margins. Major geographic zonation is maintained in the open ocean by hydrographic anomalies, affecting primary production and overall ecology, such as the frontal zones of the antarctic and arctic, the equatorial regions and zones of coastal upwelling. Here the slopes of isopycnal surfaces are generally much steeper than in the interior (i.e. about 10^{-2} as against 10^{-4}) and they may outcrop at the sea surface. The enriched nutrients in these regions often sustain high biomass. Substantial fisheries occur where coastal upwelling takes place in low latitudes (e.g. off Peru, Senegal and in the Arabian Sea), although in such systems material is brought up from only some few hundred metres depth. In contrast, at the sub-antarctic and sub-arctic fronts, the effects originate from much deeper to sustain around the sub-antarctic, for instance, vast stocks of krill which are a potentially exploitable resource. Here, too, the effect of low temperatures extending unusually far up the slope may raise the upper limit of the vertical range of bottom-living organisms. There is no evidence of large-scale outcropping of abyssal organisms, however, in near-surface waters. In addition, localized zonation in the horizontal field of the open ocean occurs when mesoscale eddies are spawned off western boundary currents, for example. In such cases midwater organisms are transported often beyond their normal range.

Apart from the oceanic fisheries for tuna and marlin, mentioned
earlier, well over 90% of the world catch of marine fish originates from
the shallow continental seas. While this includes coastal upwelling
resources, such stocks are nevertheless beyond the direct influence of
abyssal circulation. Widespread exploitation of additional pelagic
oceanic resources is unlikely on the basis of the generally low density
of any potential quarry. Likely sustainable yields of demersal oceanic
stocks are small relative to those on the continental margin. For
example, consider the area of the North Atlantic (about 4×10^{13} m^2)
with a total benthic input of about 10 g biomass m^{-2} a^{-1}. This is
4×10^8 t a^{-1} and, on the basis that only a quarter of this is
available and unutilized by the benthos and it is separated by two
trophic levels from any exploitable resource, the natural production
could only be expected to be about 1 million t a^{-1} with a sustainable
yield of, say, 250 000 t a^{-1}. Even the world ocean total sustainable
yield of deep ocean detritally maintained fish is therefore probably no
more than a few million tonnes per year.

The most likely food chains resulting in the exposure of the human
population will therefore arise in coastal waters, where the vast
majority of marine food products are harvested, and for which
considerable reliance would therefore have to be placed on model
predictions of coastal water contaminant concentrations. In these cases
the application of concentration factor data would be appropriate.
Unfortunately, direct comparisons of different model predictions are not
yet available, but it can readily be envisaged that, for example, for a
highly soluble radionuclide, such as ^{99}Tc, direct food chain transfer
is unlikely to produce a higher concentration in a predatory fish caught
in surface coastal waters than that attained, by direct adsorption, by
coastal water algae because of the latter's ability to concentrate this
nuclide (Pentreath et al., 1980). The relative importance of various
pathways will clearly be governed by the combinations of type of food
eaten by man (fish, shellfish, algae), the quantities of each which are
eaten, and the affinities of each type of foodstuff for the substance of
concern. It is therefore worth recalling that all macrophytic algae live
in waters of <150 m depth (and algae frequently have the highest
concentration factors for many elements), that the majority of shellfish
(molluscs and crustaceans) liable to be consumed at a sustained rate will
also be taken from coastal waters, and that neither of these two
foodstuffs will be linked by a predator/prey food chain to the deep
ocean. The only food chain pathway of concern, therefore, is that which
results in the consumption of fish. Fish are eaten in the greatest
quantities but tend to have the lowest concentration factors for most
elements and other potential contaminants -- with the notable exceptions
of Cs and Hg. This is partly the result of the fact that a large
fraction of their body weight consists of muscle. The general importance
of relating organ concentrations to those of whole body, and to food
eaten, has been discussed elsewhere (Pentreath, 1977) but is a factor
which should be taken into account. In view of the above, the inference
is that food chains resulting in the exposure of human population, along
critical pathways -- for dose-limit calculations -- are most likely to
arise in coastal waters. A biological need in "modelling" this is the
procurement of better concentration factor data applicable to marine
species being consumed from coastal water fisheries.

Appendix III

QUANTITATIVE ESTIMATES OF VARIOUS BIOLOGICAL PROCESSES

Biological mechanisms have often been suggested for both the rapid or intense localized transport and the large-scale bulk transport of substances from a deep-sea dump site back to man or his food chain. In this Section several biological processes are examined in order to provide a basis for deciding when and/or how they must be included in models for estimating dumping hazards. The mass of organisms in sediment and water relative to their surroundings as discussed in Section 3.3 of the main text is very small. For a 1000 m water column and underlying sediment, the total biomass could amount to 3 g m^{-2} compared with 10^3 t m^{-2} of water. The greater part of any contaminant will therefore reside in the water unless concentration factors exceed 3×10^8. It is for this reason that in comparison to physical ones, biological transport mechanisms are usually unimportant for large time- and space-scales but may need to be carefully analysed when considering certain critical paths or local problems. Of course, ultimately biological processes control the amount of substances entering man's food chain but usually this will occur after the substance has been transported great distances from the dump site.

1. CALCULATIONS OF MAXIMUM BIOLOGICAL TRANSPORT

1.1. Transport by swimming fish

Consider an abyssal fish population which accumulates a contaminant, and transports it by swimming. An assumed standing stock of 0.3 g m^{-2}, concentrated within 10 m of the seabed, yields a ratio of biomass to water occupied of 3×10^{-8} (weight/weight). Thus even if the fish swim a hundred times faster than the water moves (1 m s^{-1} compared with 1 cm s^{-1}, for instance) the amount of contaminant transported by fish is less than that transported by water, unless the concentration factor for the contaminant exceeds 3×10^5. Even with this concentration most of the contaminant would remain in the fish, and not be released to the environment.

1.2. Potential vertical exchange based on carbon budgets

An alternative approach to improve a general estimate of potential vertical export is to utilize new data on the input of organic energy to deep-ocean ecosystems to obtain approximations of total possible production. A simplified representation of a benthic boundary layer food chain can be constructed to aid in interpreting what parts of that production would be exported (Figure A.III-1).

The flow of energy (sinking particulate organic carbon, POC) that maintains life in the deep sea has been measured using particle traps moored near bottom in both the Atlantic and Pacific Oceans at all but the greatest depths. These POC measurements include the 4 to 5 km depth

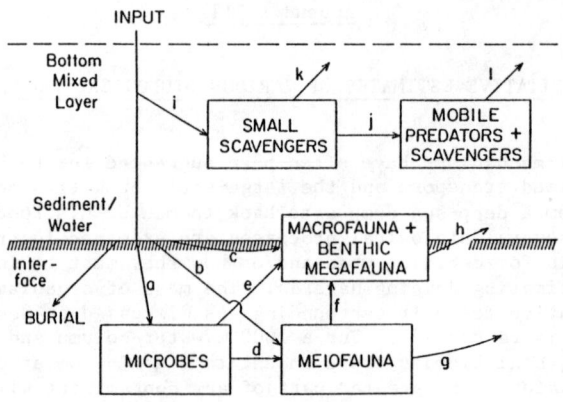

INPUT

Bottom Mixed Layer

SMALL SCAVENGERS

MOBILE PREDATORS + SCAVENGERS

Sediment/ Water

MACROFAUNA + BENTHIC MEGAFAUNA

Inter-face

BURIAL

MICROBES

MEIOFAUNA

FIG. A.III–1.   Potential export = (k+l+h+g) × concentration factor.

range that encompasses present or potential deep-ocean dump sites (Convention on the Prevention of Pollution by the Dumping of Wastes and Other Matter, 1972).   This flux is the input term of the model, and it ranges in value from several grams carbon m^{-2} a^{-1} on the upper continental rise to less than 0.5 g carbon m^{-2} a^{-1} on deep abyssal plains.   A portion of the annual supply (1 - 20%) of this is buried as the sediment accumulates.   The remainder is utilized by bacteria, meiofauna and macrofauna that live in or on the sediment, and by benthic and benthopelagic fishes and invertebrates, which live on and above the sediment.   A group of scavengers completes the food web between the fish and the small detrital particles raining slowly to the sea floor.   This identification of five animal groups is based on the size of the organisms and how they are sampled, not necessarily on their roles in the ecosystem.   The separation does allow some assumptions to be made about food chain relationships and reproductive strategies based on size. Given this information one can infer how mass may be transferred through or out of the system; knowledge of reproduction and development allows assumptions to be made on what groups will export eggs, larvae or juvenile stages out of the system.

The input of particulate organic matter has two fates: burial or biological consumption.   The biological utilization can be divided into consumption by the various groups of organisms at rates proportional both to their individual and collective biomass, represented by first-order, linear functions.   The organisms consume each other in a manner related to their sizes, e.g., bacteria are eaten by meiofauna which are eaten by macrofauna, the latter also consuming the bacteria, etc.; small organisms generally have relatively higher turnover rates than larger ones.   These rates can also be assumed to be regulated by the proportion of the biomass in the community comprising each group.

The major fraction of biologically assimilated energy will be used in respiratory conversion to CO_2.   It is generally accepted that at least 50% and more on the order of 90% of the energy consumed suffers this loss, the remainder going to production.   Arbitrarily therefore the loss of CO_2 has been assumed to be 70 to 80% of the input to all five groups of consumer organisms.   What is left will be consumed by a higher trophic

level or be exported as eggs. Although the details of trophic
interactions are poorly known, by rearranging terms it becomes obvious
that input less burial and total CO_2 production equals the total export
term.

Input, burial and total respiration by macrofauna, meiofauna and
bacteria are known from a range of depths in a few areas, including the
Northwest Atlantic (Rowe and Gardner, 1979; Hinga et al., 1979). In each
of 4 Northwest Atlantic regions at depths of 2.2 to 5.2 km an imbalance
in the carbon budget was observed (Rowe and Gardner, 1979) while Hinga et
al. (1979) observed an imbalance at depths greater than 3500 m. Rowe and
Gardner (1979) found that the imbalance had a magnitude of about 25% of
the input, comparable to the respiration. Other sources of respiration
exist however that were not measured, e.g. larger organisms and an
intermediate benthopelagic trophic level, which means "possible" export
would be somewhat smaller than an amount inferred only by difference
between input less burial and infaunal respiration. The imprecision of
current estimates of supply and loss should, however, be emphasized.

Several important differences can be seen between this model and that
of Robinson and Mullin (1981). The energy is used primarily for
respiration, which the latter authors concede is important. While
Robinson and Mullin considered that all production could be exported, the
calculations presented here assume that all bacterial production is
consumed by meiofauna and macrofauna. Megafauna, which include roaming
and sessile large invertebrates on the sediment surface, are probably
important consumers of all three smaller types of organisms. They are
known to have reproductive stages that would be exported. Using
respiration data of Smith (1978) and biomass data from Haedrich and Rowe
(1977), Rowe (1981) concluded that the total carbon consumption of the
larger organisms relative to the smaller ones was small, but this needs
further evaluation when more comprehensive data become available.

The above model can be used to calculate potential transport in the
western North Atlantic, from the Continental Shelf to the Hatteras
Abyssal Plain. The regressions of particulate organic carbon (POC)
burial and respiration on sedimentation of organic carbon (Rowe, 1981)
cross the 4 km isobath, the upper depth limit defined by the London
Dumping Convention. At that depth, the three critical variables can be
estimated, the difference being potential export:

$$\text{Flux} = 2 \text{ g carbon m}^{-2}\text{a}^{-1} = \underset{\text{burial}}{0.5 \text{ g carbon m}^{-2}\text{a}^{-1}} + \underset{\text{respiration}}{1 \text{ g carbon m}^{-2}\text{a}^{-1}}$$

+ unmeasured respiration by fishes and intermediate detritivores
+ export

where the unmeasured respiration and export equals 0.5 g carbon $m^{-2} a^{-1}$.

This analysis indicates that at most 0.5 g carbon $m^{-2} a^{-1}$ is available
for export by living organisms per year. Assuming as for Table A.III-I
that 1 g carbon is contained in 24 g wet weight of all organisms, this
corresponds to an annual production of 12 g carbon $m^{-2} a^{-1}$ of living
organisms. An extreme estimate of average vertical transport of a
contaminant by biological matter may be obtained by assuming that all
this biomass is transported to the surface carrying a contaminant
characterized by a concentration factor γ. This may be compared with the

ESTIMATES OF WET WEIGHT BIOMASS (g m^{-2}) FOR VARIOUS SIZE GROUPS OF
BENTHIC ORGANISMS IN SEDIMENT (TO A FEW cm DEPTH) ON AND ABOVE
THE BOTTOM AT ABYSSAL DEPTHS $>$4000 m

Data are from Atlantic and Pacific sites which vary in
levels of primary production. 1 g carbon is assumed to be contained
in 24 g wet weight for all organisms except for bacteria

Group	Biomass (g m^{-2})	Carbon Equivalent (mg carbon m^{-2})	Reference
Bacteria	0.2-0.6	20-60	Watson et al. 1977; Deming, 1981; Williams and Carlucci, 1976
Meiofauna ($>$50 m)	0.05-0.5	2-20	Thiel, 1975; Dinet (pers. comm.)
Macrofauna ($>$ 250 m) (excl. foraminifera)	0.01-10.0	4-400	Rowe, et al. 1974; Haedrich and Rowe, 1977; Rice (pers. comm.); Rowe, 1983
Megafauna (epifauna) ($>$ 1 cm)	0.02-1.0	0.8-40	Haedrich et al. 1980;
(nekton)	0.01-0.1	0.4- 4	Wishner, 1980; Sibuet (pers. comm.)
Fish (benthic and benthopelagic)	0.02-1	1-40	Merrett (unpubl. data)
(abyssopelagic integral 4000-5000 m)	0.01	0.4	Angel and Baker, 1982

net physical transport by a typical average upwelling velocity of
$w = 10^{-5}$ cm s^{-1} or 3 m a^{-1}. One obtains

$$\frac{\text{average contaminant transfer by biological matter}}{\text{average contaminant transfer by physical upwelling}} =$$

$$\frac{\gamma C \times 12 \text{ g m}^{-2} \text{ a}^{-1}}{C \times 3 \text{ m a}^{-1} \times 10^6 \text{ g m}^{-3}} = 4 \times 10^{-6} \gamma$$

Thus, even for a relatively large concentration factor ($\gamma = 10^3$ to 10^4) the average vertical biological transport will be small compared to the physical transport. This calculation may be compared with that of Robinson and Mullin (1981) who came to the same conclusion. Instead of estimating production of biomass from carbon budget considerations, however, they used the deep-sea biomass and estimates of turnover rates to estimate production.

Although this analysis implies physical transport of a contaminant is more important than that by organisms, it should be noted that it only applies to yearly and basin-wide averages. Concentration of the biological transport in space or time could serve to make it relatively more important. One example of such an effect is discussed in the next Section. Biological transport is also of course important when the organisms are involved in a critical pathway to man or his food chain, no matter how small a contribution it makes to the total transfer of the contaminant (see Appendix X).

1.3.   Release of reproduction products

In the previous Section the average vertical transport by biological matter was compared to that by average physical transport processes and found to be small. The question arises, however, as to whether or not the biological transport might be more important if it was concentrated in a local area, say, the dump site. One might suppose, for example, that a dominant (abundant, several grams m^{-2}) organism with wide geographic (at least several hundreds of kilometres) distribution, perhaps including continental slope regions, aggregates yearly to spawn. One example could be abyssal rattail fishes, although almost nothing is known of their spawning habits. Another could be the widespread slope-dwelling eel, Synaphobranchus kaupi, which in the Atlantic spawns in the Sargasso Sea region. If spawning aggregation occurs in the presence of maximum concentrations of contaminants, e.g. around the dump site, and if the maximum concentration in reproductive products is reached quickly, then the release of buoyant eggs might provide a significant contaminant transfer mechanism.

If the species in question spawns 1/4 of its body weight and has an average biomass of 1 g m^{-2} over an area of 10^6 km^2, then the production (P) of buoyant eggs could be 2.5×10^{11} g a^{-1}. The average concentration of the contaminant in the dump site area for a diffusive ocean is given by (see Appendix VII)

$$C_{max} = \frac{Q}{\pi (K_V K_H)^{\frac{1}{2}} r_S} \qquad \text{(A.III.1)}$$

where r_S is the radius of a distributed source (the dump site) area. If the area of spawning is greater than the dump site, the appropriate average concentration can be obtained within a factor of 2 by replacing r_S by the radius over which spawning takes place (see Appendix IV).

The total flux (F_S) of the contaminant to the surface by buoyant eggs, is given by

$$F_S = \gamma P C_{max} \rho_B^{-1} \qquad \text{(A.III.2)}$$

where ρ_B is the density of the eggs. This total flux when averaged over the area from which the fish arrive will be small for long-lived contaminants compared to that caused by physical processes for the same reasons as given in Section 1.2 of this Appendix.

The flux, Equation (A.III.2), constitutes a source to the surface water that itself is dispersed, as is the bottom source, according to Equation (A.III.1). The resulting maximum surface concentration would then be

$$C_S \quad = \quad \frac{F_S}{\pi (K_V K_H)^{\frac{1}{2}} r_S} \quad = \quad \frac{\gamma Q P}{\rho_B \pi^2 K_V K_H r_S^2} \tag{A.III.3}$$

assuming the surface and bottom diffusivities are equal. For contaminants whose decay over the rise time for the eggs is small, one can compare Equation (A.III.3) with the concentration $C = Q(\lambda V)^{-1}$ for the average concentration in the far field. The ratio A of C_S to C is

$$A = \frac{\gamma \lambda P V}{\rho_B \pi^2 K_V K_H r_S^2} \tag{A.III.4}$$

Taking P as above, $V = 10^{17}$ m^3, $K_V = 10^{-4}$ m^{2}s^{-1}, $K_H = 10^2$ m^{2}s^{-1}, $\rho_B \approx 1$ and the area of the dump site to be 10^4 km^2, one obtains

$$A = 3 \lambda \gamma \times 10^{-6} \tag{A.III.5}$$

where the units of λ are s^{-1}. For a long-lived element such as plutonium ($\lambda \approx 9 \times 10^{-13}$ s^{-1}), A is small as long as γ is less than 10^5. For shorter-lived elements, however, this will not be the case and the concentration arising from buoyant eggs could be significant in comparison to that arising from average physical processes. This is of course not surprising since for short-lived contaminants the short transit time for the proposed biological flux allows some of the contaminant to reach the surface before it decays. The calculation as performed is also for the continuous release of buoyant products. A pulsed release over a short period would give higher concentrations temporarily.

Whether such a biological process would ever be significant depends more on the value of the flux F_S and whether or not a critical pathway is involved than on the relative transports of the contaminant due to different processes. In this context one should note that for both short- and long-lived material the concentration in living organisms near the source will always be greater than at the surface.

2. FERTILIZATION EFFECTS

Standing stock and biomass of the deep-sea fauna decrease exponentially with depth and it is presumed that this is caused by limited food supplies. If a carcass of a dead fish is put on the sea floor, scavengers accumulate around it and consume it, often very quickly. If wood is introduced it rapidly becomes infested with boring molluscs. The waste matter from the borers supports a detritus-eating community in close proximity to the wood as well. Such observations suggest that the introduction of organic matter into the deep sea could significantly increase standing stocks. Experiments to test the rate of succession of deep-sea communities and how they respond to such alien inputs are inconclusive so far, but attention should be paid to some situations where the biological transport processes could be altered by the introduction of alien organic supplies.

In general, solid wastes that would be deposited in dump sites would not contain enough organic matter to present such a problem. Some important exceptions, however, might occur. Areas of contaminated soils that are removed can contain many different components, including all kinds of organic detritus and flora, including trees, shrubs, etc. Placing such materials in dump sites could appreciably fertilize a deep-sea area, altering species diversity and abundance of organisms. Attention should be paid to how non-radioactive waste components will alter biological pathways. Organic wastes which could include experimental animals, and other laboratory materials, may need segregation from other materials whose entrance into a food chain must be minimized.

3. THE ARTIFICIAL REEF EFFECT

A possible consequence of deep-sea dumping is the formation of an artificial reef effect. Structures placed on the bottom on shallow, flat continental shelves improve fishing by attracting mobile carnivores and scavengers, by serving as protective refuge for a variety of species. If this "attraction" occurs at deep-sea dump sites, it could increase the potential biological accumulation of a substance by increasing total biomass exposed to radionuclide release. The presumptions about the rates of deep-sea biological transfer may not be applicable in such cases. Total biomass might increase and species composition could be altered to be composed primarily of adhering species living directly on and attached to the waste canisters. How this occurs will depend on the environment at the location and the physical configuration of the waste containers (barrels, submarine hulls, etc.). This will put organisms in direct, close contact with the substance being released.

When reefs are constructed in shallow water to improve fishing it is questionable whether or not they actually increase total productivity from photosynthesis up through the food chain associated with the structure. They could be just concentrating fish. The argument that total production increases is usually based on the effects that the added structure has on flow and turbulence around it. Mixing is increased, as is hard surface area, and both contribute to higher planktonic and attached algal production.

In the deep sea, where no photosynthesis is possible, there seems to be no possibility that higher total production will occur unless altered flow fields accentuate deposition of organic detritus. Attached filter feeders are often common in the deep sea where predictable currents and hard substrates are available. These include principally hydroids, bryozoans, stalked barnacles, glass sponges and crinoids. These, as well as mobile animals such as brittle stars, starfish and urchins, would be expected to be attracted to a dump, in the same way as occurs on a shallow water artificial reef. Carnivorous and scavenging fishes, in this case the rattails, brotulids, rays, etc., will probably be attracted to the increased food supply.

The question of how much this will contribute to biological concentration and to transfer of substances probably depends on the degree to which the dump differs in biomass and production from the surrounding deep sea. Most bottom water is low in suspended particulate organic matter, with concentrations on the order of 5 - 20 μg carbon L^{-1}. Total suspended matter (TSM) is often increased just above the bottom due to resuspension and this will include POC, but the concentration of POC will be low, reflecting the bottom rather than settling pelagic organic debris. Because this energy supply (POC) is universally so low in the deep sea, it seems highly unlikely that any concentrating factors related to flow field, erosion and deposition around a structure will appreciably increase biomass and production.

Even though the net production over a broad area would remain the same, increases in biomass and production in localized areas as well as a change in species composition might be expected. This should be no particular cause for alarm. This alteration should be documented with the idea in mind that new pathways might be introduced that would not have been found in a predumping site assessment.

A further ramification is that, because of the short time- and space-scales involved, any physical, geochemical and biological transport models that might be required for the extreme near field would have to evolve as a dump site fills up with debris. Bottom roughness, turbulence, erosion and deposition, and the biota will all change with the addition of more and more material. How these change will primarily depend on waste form and dumping strategies, as well as the potential for change in the environment.

4. CALCULATION OF THE HORIZONTAL AREA REQUIRED TO SUPPORT DEEP-SEA FISH PRODUCTION

In assessing the impact of a contaminant via a fishery, it is useful to have an estimate of the area of sea floor capable of supporting a fishery of any given magnitude. The biomass per unit area of organisms in the deep sea is much lower, perhaps by a factor of a thousand, than that on the continental shelves (see Appendix II). Although the absolute values of production are not yet known, it is reasonable to assume that these are correspondingly small, especially since rates of growth and respiration are also believed to be much lower than in shallow water. Thus, the area of deep-sea floor required to support a fishery will be very much greater than that in shallow water. This area can be estimated more specifically from a knowledge of the input of organic matter to the deep sea, from comparison with an estimate of the biomass generally believed to be required to support a continental shelf fishery and from some ecological generalizations based on the known estimates of deep-sea fish stock size and respiration rates.

As a reference calculation one may estimate the area required to support a minimum credible fishery, sustaining a yield of only one fish per day weighing 600 g (200 kg a^{-1}). Such a yield might (allowing for weight lost during filleting, etc.) be consumed by only one enthusiastic consumer (eating 300 g d^{-1}) or a small group of more moderate consumers (30 people x 10 g d^{-1}). A common rule of thumb in fisheries is that the sustainable yield from a pristine biomass (B) is roughly 0.25 x M x B, where M is the natural mortality rate and 0.25 is an empirical constant. A value of M = 0.1 a^{-1} would be considered low in shelf fisheries, but may be appropriate for the deep sea where turnover rates are lower. The pristine biomass required to sustain the given annual yield may thus be taken to be about 40 times the yield, or 8000 kg a^{-1}.

As noted in Section 1.2 of this Appendix, the major source of food for deep-sea communities is particulate organic detritus. The potentially exploitable fishes are not detritovores, but predators, so at least one trophic level must exist between the organic detrital supply and such a fish population (Figure A.III-1). An assumption (based on shallow water fishes) that the fish population consumes about 1% of its weight per day implies a consumption of 80 kg d^{-1}. Taking the energy conversion efficiency of the intervening trophic level to be 10%, the detrital food supply required is 800 kg d^{-1} or about 300 000 kg a^{-1}. It has been estimated in Section 1.2 that of the energy flux to the deep sea only about 20% is available for production, that is, most of it is either utilized by the benthos or is accumulated on the bottom in a presumably refractory form. The amount available for production would be no more than 2.0 g m^{-2} a^{-1} of fresh organic matter or about 0.1 g carbon m^{-2} a^{-1}. Comparison of this supply with a required 300 000 kg a^{-1} ingestion suggests that on the order of 1.5×10^8 m^2 is needed to support the limited fishery (200 kg a^{-1}).

Growth rates and respiration rates per unit of biomass are believed to be much lower in the deep sea than in shallow water, and therefore the above estimate based on shallow species may not be very accurate. An alternative approach for estimating the ingestion rate for the population would be to base it on measures of respiration and then extrapolate that to total consumption based on possible trophic efficiency of the population. Respiration of the rattail <u>Coryphaenoides armatus</u>, using data from the Northwest Atlantic as an example, is in the order of 3 ml O_2 kg^{-1} h^{-1} (Smith, 1978). Taking a reasonably high value of 0.5 g m^{-1} (Table A.III-I)) for the fish biomass at greater than 4000 m depth, respiration would amount to:

$$(3.0 \text{ ml } O_2 \text{ kg}^{-1} \text{ h}^{-1}) \ (0.5 \text{ g m}^{-2}) \approx 13 \text{ ml } O_2 \text{ m}^{-2} \text{ a}^{-1}$$

This may be directly converted to give a respiration of 0.006 g carbon m^{-2} a^{-1}. It is thought that production of biomass by deep-sea fish lies between 5 and 20 per cent of respiration. Of this one might guess that no more than one-half is available as sustainable yield (although the empirical relationship based on shelf fisheries given above uses one-quarter). Using these estimates one obtains the sustainable yield of fish to be 0.00015 - 0.0006 g carbon m^{-2} a^{-1} or 0.0036 - 0.0014 g m^{-2} a^{-1} of wet weight. Thus, the total area required to sustain the limited fishery of 200 kg a^{-1}, is found to be $1.4 - 5.6 \times 10^7$ m^2, an area slightly less than estimated above.

Both the above estimates rely on assumptions about the nature of potential deep-sea fisheries. The first extrapolates to the deep-sea empirical formulae for sustainable yield and consumption obtained for surface fisheries and uses the still uncertain estimates of the flux of organic matter to the deep sea. The second uses equally uncertain estimates of the respiration and biomass of deep-sea fish and requires assumptions to be made regarding the ratios between respiration, production and sustainable yield. None of the steps of the two calculations are, however, inconsistent with what is known of the deep-sea ecosystem. The calculations also are for an equilibrium fishery in a particular location and are not appropriate for estimating the initial yield from a given area or the yield from a moving fishery exploiting virgin stocks. The question of enhanced fisheries in a dump site has been addressed elsewhere in this Appendix.

Appendix IV

A SIMPLE FINITE OCEAN DIFFUSIVE MODEL

If a tracer with decay rate λ is released at a rate Q into an ocean of volume V, with no removal at its boundaries, the average concentration will approach $Q(\lambda V)^{-1}$ after a time which is long compared with λ^{-1}. Departures from this average (especially near the source), and the time-development of the final steady state, may be calculated from a full three-dimensional model incorporating diffusion, advection, and any other relevant processes.

As an aid to the interpretation of the results from such a three-dimensional model (Shepherd, 1976), a simple analytical, steady, one-dimensional diffusion model was proposed by Shepherd and incorporated into Document IAEA-210 as Appendix III. In particular, this model gave the bottom concentration relative to the well-mixed average as a function of D/Δ, where D is the ocean depth and $\Delta = (K_V/\lambda)^{\frac{1}{2}}$ is the characteristic length scale of vertical inhomogeneity, with K_V being the vertical diffusivity. For $D \gg \Delta$, Shepherd showed that the relative bottom concentration tends to D/Δ, with relative concentrations greater than one over a depth of order $\Delta \ln(D/\Delta)$, whereas for $D \ll \Delta$ the relative concentration tends to 1 everywhere.

The purpose of this appendix is to make the simple extension of Shepherd's model to the three-dimensional case, in order to draw attention to the much higher near-source concentrations than predicted by the one-dimensional model.

If K_H, K_V are the (constant) horizontal and vertical diffusivities, the equation for the scaled concentration $\widehat{C}$ is

$$\frac{K}{r} \frac{d^2}{dr^2} (r\widehat{C}) - \lambda\widehat{C} = 0 \qquad\qquad (A.IV.1)$$

where $r = [(K/K_H)\cdot(x^2+y^2) + (K/K_V)z^2]^{\frac{1}{2}}$ is the radial distance, from the source at $r = 0$, in a scaled ocean and the diffusivity K is arbitrary.

The boundary conditions on the scaled concentration, $\widehat{C} = C\, K_H K_V^{\frac{1}{2}}/K^{3/2}$, are

$$-2\,\pi r^2 K\, d\widehat{C}/dr = Q \qquad \text{at} \qquad r = 0 \qquad\qquad (A.IV.2)$$

and zero normal flux across any boundary. To permit a simple solution of Equation (A.IV.1) that depends on the spatial co-ordinates only through r, we pretend that in terms of the scaled co-ordinate r the ocean is hemispherical and of radius R, and will argue later that our main result is independent of this absurd assumption. The second boundary condition on C is then:

$$d\widehat{C}/dr = 0 \qquad \text{at} \qquad r = R \qquad\qquad (A.IV.3)$$

The solution of Equation (A.IV.1) subject to Equations (A.IV.2) and (A.IV.3) is then

$$\hat{C} = \frac{Q}{2\pi Kr}\ [\cosh\frac{r}{\Delta} + \frac{1-\epsilon\ \tanh\ \epsilon}{\epsilon-\tanh\ \epsilon}\ \sinh\frac{r}{\Delta}\ ] \qquad\qquad (A.IV.4)$$

where, as before, $\Delta = (K/\lambda)^{\frac{1}{2}}$ and we define $\epsilon = R/\Delta$. Relative to the "well-mixed" concentration $C_0 = Q(2\pi R^3\lambda/3)^{-1}$, we have

$$\hat{C}/C_0 = \epsilon^3(3r/\Delta)^{-1}\ [\cosh(r/\Delta) + (1-\epsilon\ \tanh\ \epsilon)(\epsilon-\tanh\ \epsilon)^{-1}\sinh(r/\Delta)] \quad (A.IV.5)$$

For $\epsilon >> 1$ (i.e. for the decay time λ^{-1} much less than the oceanic mixing time R^2/K), we have

$$\hat{C}/C_0 \simeq \epsilon^3(3r/\Delta)^{-1}e^{-(r/\Delta)} \qquad\qquad (A.IV.6)$$

or $\qquad \hat{C} \simeq Q(2\pi Kr)^{-1}e^{-(r/\Delta)} \qquad\qquad (A.IV.7)$

This is the distribution for a source in an infinite half-space; unlike the one-dimensional situation the concentration has a singularity at the origin. We note also that outside the near-source singularity the e-folding scale for the concentration is $\Delta = (K/\lambda)^{\frac{1}{2}}$. The concentration falls off less rapidly at $r = \Delta$, than would be estimated from $\exp(-\lambda t)$ with $t = r^2/K$ the time to diffuse to distance r. This can be interpreted as an increase of gradient, and hence of diffusion, due to the decay.

In dimensional terms Equation (A.IV.7) may be written as

$$C = \hat{C}K^{3/2}K_H^{-1}K_V^{-1/2} = Q(2\pi)^{-1}[(K_HK_V)(x^2+y^2) + K_H^2z^2]^{-1/2}$$

$$x\ \exp\left|-[\ \lambda K_H^{-1}(x^2+y^2)+ \lambda K_V^{-1}z^2]^{1/2}\right| \qquad (A.IV.8)$$

The distribution for $\epsilon << 1$ (i.e. long decay time compared with the oceanic mixing time) is of rather more interest. In the one-dimensional problem the relative concentration tends to 1 everywhere, but in three dimensions

$$C/C_0 \approx 1 + \epsilon^3(3r/\Delta)^{-1} \qquad\qquad (A.IV.9)$$

Thus the relative concentration is close to 1 except within a source region. In fact Equation (A.IV.9) may be written

$$\hat{C} = Q(\lambda\hat{V})^{-1} + Q(2\pi Kr)^{-1} \qquad\qquad (A.IV.10)$$

where the scaled volume $\hat{V} = 2\pi R^3/3$. In dimensional terms this is

$$C = Q(\lambda V)^{-1} + Q(2\pi)^{-1}[(K_HK_V)(x^2+y^2)+K_H^2z^2]^{-1/2} \qquad (A.IV.11)$$

These are clearly general results, independent of the shape of the ocean, showing a source distribution superposed on top of a uniform distribution. We note that the amount of contaminant associated with the source solution is of the order of Q times the mixing time, which is, by assumption, much less than the required inventory of Q times the decay time λ^{-1}.

Equations (A.IV.10) and (A.IV.11) show a source region within which the concentration is significantly greater than in the far field. If we define the source region by the concentration contour C = 2 x the far field, or average, then it is ellipsoidal with horizontal scale $\lambda V(2\pi)^{-1}(K_H K_V)^{-\frac{1}{2}}$ and vertical scale $\lambda V(2\pi)^{-1}K_H^{-1}$. For plutonium with $\lambda = 9 \times 10^{-13}$ s^{-1}, if we take an ocean of radius 3 x 10^3 km, and depth 4 km, and assume $K_H \simeq 10^2$ m^2 s^{-1} and $K_V = 10^{-4}$ m^2 s^{-1}, we obtain horizontal and vertical scales of 150 km and 150 m.

The point-source solution in Equation (A.IV.11) could be used to estimate the near-field concentration, though in the extreme near field, i.e. within the benthic boundary layer, one would use the higher value of K_V that would be appropriate.

In the presence of a mean flow with speed U, lateral diffusion will dominate advection out to a scale of order K_H/U, and the source-like solution above will be valid within this distance of the source. For typical values of K_H and U this distance is of order 10-100 km.

In a time-dependent model the near-source concentration field will be established quickly for a switched-on source, and maintain its values over the much longer time taken for the rest of the ocean to reach equilibrium concentration. The time-scale for the near-field solution to changes in source strength is in fact just the time taken to mix across the near-field region, that is, the (length-scale)^2/K. This is of the order of one year both laterally and vertically.

The shortcomings of this model hardly need listing, but it does show that a one-dimensional model can be misleading, and can be replaced by an equally simple three-dimensional model.

Appendix V

PARAMETRIZATION OF BOUNDARY SCAVENGING PROCESSES

As stated in Appendix IV, a substance, whose only removal mechanism is radioactive decay at rate λ, when released at a rate Q into an ocean of volume V reaches a steady-state concentration of $Q(\lambda V)^{-1}$. However, other mechanisms such as scavenging by particles, diffusion into the bottom, and burial may remove the substance. In this Section the parametrization of such removal processes is discussed. The details of their effect on the interior distribution of the substance are discussed elsewhere.

Much can be learned by examining in detail the possible partition of a substance between the water phase, particles in the water column, and the sediments, if time is given to reach a steady-state situation and the concentration of a substance associated with particles and sediments is in equilibrium with the water concentration.

First let us consider sorption by suspended particles and biota. Typically the average level of suspended particulate matter is $10 \, \mu g \, kg^{-1}$ ($f \sim 10^{-8}$). Thus for $K_d < 10^8$, the concentration per unit volume of a substance carried on particulate matter, C_p, is less than in the water phase, C; and except for highly reactive elements very much less. However, as is noted in Appendices VI and IX, the flux of a substance by falling particles may be of first-order importance.

The load of a substance associated with suspended particles may also be compared to the thickness of an equivalent layer of sediment. Indeed using the values given above, the total particulate load in an ocean of 4000 m is only equal to a layer of sediment 0.004 m thick. Similarly, one may note that a typical standing crop of biomass of $1 \, g \, m^{-2}$ would hold the same quantity of a substance as $10^{-4} \, \gamma/K_d$ cm of sediment. This is less than 1 cm even for a high concentration factor ($\gamma \sim 10^4$) and low K_d(~ 1). The amount of a substance carried by either suspended particulates or biota will therefore be negligible compared with that on bottom sediments or the water column and cannot significantly affect the ultimate budget (mass distribution) of a substance in this context.

Now let us consider the amount of a substance that may be contained in the sediments by three different processes.

(1)   A superficial layer of sediments (depth h) that is rapidly turned over and mixed (e.g. by bioturbation), can hold as much of a substance, taking into account the sorptive capacity of the sediments, as a layer of water of depth $h^* = h(1 - f' + f'K_d)$, where f' is the fraction of the volume occupied by sediment. Thus, if the water is also well mixed and in equilibrium with the sediments, the effective ocean volume is increased from V to $V(1 + h^*/H)$ and the specific concentration is reduced to $[\lambda V(1 + h^*/H)]^{-1}$. If the ocean is not well mixed, the effect is qualitatively the same but quantitatively different.

(2)   If there is net accumulation of sediments, the substance will be removed from contact with the water column by newly arriving sediment. This process will be referred to as burial. Since it decays away in an average time of $1/\lambda$, the thickness of sediment that remains contaminated is given by w_s/λ and it will contain as much of the substance as a layer of water of thickness

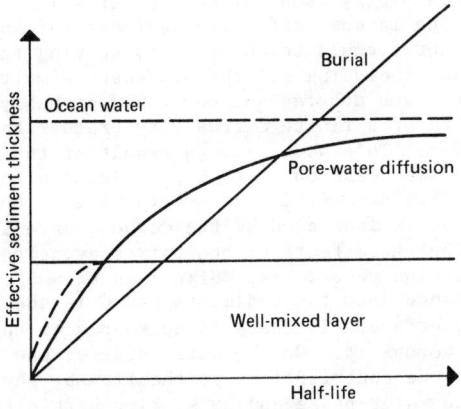

FIG. A.V-1. Schematic diagram of effective sediment thickness.

$w_s((1-f')+f'K_d)/\lambda$ where w_s is the net accumulation rate of
sediments. For reactive sediments this is approximately
$w_s f'K_d/\lambda$. One should note that $f'w_s$ may differ by a factor $\delta < 1$
from the flux of particulate matter to the bottom, $|fw_p|$, if
the organic component dissolves on reaching the bottom. Two
possibilities then occur. Firstly, as the organic matter
dissolves, the substance it carries can be retained and bound to
the remaining sediments, effectively increasing the sedimentary
value of K_d for its water column value; or, secondly, the
substance can be released to the pore water, a possibility
included in the model of Appendix IX.

(3)   A substance may also be carried into the underlying sediments by
      pore-water diffusion. Since it also decays there this leads to
      an exponentially declining profile in the sediments with a scale
      length $(K_{eff}/\lambda)^{\frac{1}{2}}$ where K_{eff} is the effective pore-water
      diffusion coefficient for a reactive substance. To a first
      approximation, this may be taken as K_{pw}/K_d, where K_{pw} is
      the pore-water diffusivity allowing for porosity and
      tortuosity. Thus the effective sediment thickness for diffusive
      removal is $h_{diff} = (K_{pw}/\lambda K_d)^{\frac{1}{2}}$.

   The nature of these effective sediment thicknesses is illustrated
graphically in Figure A.V-1. The well-mixed layer and the ocean itself
are represented by straight horizontal lines corresponding to constant
thicknesses (with respect to variations of half-life). Their relative
position depends primarily on the value of the distribution coefficient,
K_d. The thickness for net sediment accumulation increases in
proportion to the radioactive half-life, and is therefore always the
largest at very long half-life. That for the pore-water diffusion
process is a parabola, since it is proportional to $(1/\lambda)^{\frac{1}{2}}$.
   The relative position of the lines depends not only on λ but on the
values of K_d, K_{pw}, w_s, δ , f', and various other environmental
variables. If all the latter were known there would of course only be

one curve which would pretty much follow the curve in Figure A.V-1 that is dominant (gives the largest effective sediment thickness) for each value of λ. This curve could be obtained by solving the appropriate sedimentary equations including all the processes simultaneously. In this case for short-lived substances, one would probably also want to allow the possibility of a finite diffusivity (rather than infinite) in the bioturbated layer. This would give a result of the same form as for the pore-water diffusion case but with K_{eff} being replaced by the much larger diffusivity, K_b, descriptive of overturning of the layer by biota. This addition is indicated by the dashed curve in Figure A.V-1.

For parametrizing the effects of boundary scavenging processes, it is useful to use deposition velocities, defined as $V_d = F/C$, where F is the flux of a substance into the sediments and C is the concentration in the water at the interface. If there is no source at the interface, F will be continuous across it. On the water side of the interface F should include both the contributions to the flux by physical processes (e.g. mixing) in the water phase and by sinking particles. The deposition velocity has fairly general applicability for describing removal processes at boundaries by a wide range of physical and chemical processes. Problems arise however if the sediments and water are not in equilibrium, a possibility discussed in Appendix IX.

The effects of boundary scavenging processes may also be expressed in terms of removal rates; that is, contributions to λ_{eff} defined by $C = Q(\lambda_{eff}V)^{-1}$. These may be compared to the removal by radioactive decay in the water column alone at a rate λ.

In order to illustrate the utility of V_d and λ_{eff}, consider individually the three removal processes treated above in the discussion of effective sediment thickness. Although a more accurate treatment of the problem is again to treat all mechanisms simultaneously, it will be seen that this is not necessary for the purpose of identifying which processes are dominant for a substance with a particular half-life and reactivity.

(1) As given above the effect of a rapidly well-mixed superficial sediment layer is to increase the ocean volume by a factor $(1+h^*/H)$, so that

$$CQ^{-1} = (V\lambda_{eff})^{-1} = [V\lambda \; (1+h^*/H)]^{-1} \qquad (A.V.1)$$

and

$$\lambda_{mix} = \lambda_{eff}-\lambda = (h^* \lambda)/H \qquad (A.V.2)$$

where λ_{mix} is the removal rate for the well-mixed layer of sediment. Since the flux into the sediment is given by the removal per unit area, the deposition velocity is simply

$$V_d = H\lambda_{mix} = h^*\lambda = h(1-f'+f'K_d)\lambda \qquad (A.V.3)$$

(2) For burial, the deposition velocity is simply determined by the net surface flux of a substance, that is

$$V_d = w_s(1-f'+f'K_d) \qquad (A.V.4)$$

The characteristic removal rate constant for this process is

$$\lambda_{sed} = V_d/H = w_s(1-f'+f'K_d)/H \qquad (A.V.5)$$

(3) The balance between diffusion and decay in the underlying sediments causes an exponentially declining concentration profile, C_s, where

$$C_s(z) = C_s(0) \exp(-\beta z) \qquad (A.V.6)$$

and $\beta = \sqrt{\lambda/K_{eff}}$

The flux at the interface is given by $F = -K_{eff}dC_s/dz$, evaluated at $z = 0$.

The deposition velocity is then

$$V_d = F/C = K_d \sqrt{K_{eff}\lambda} \qquad (A.V.7)$$

Or, as before, taking $K_{eff} = K_{pw}/K_d$

$$V_d = \sqrt{K_d K_{pw}\lambda} \qquad (A.V.8)$$

and the effective removal rate constant

$$\lambda_{diff} = V_d/H = \sqrt{K_d K_{pw}\lambda} \, / \, H \qquad (A.V.9)$$

These results are summarized in Table A.V-1.

As mentioned above these formulae are only strictly true if a single removal mechanism is effective[3]. One important adjustment to the

---

[3] If both diffusion into the sediments, characterized by a diffusivity K_{eff}, and burial are important, the situation is also easily treated as long as one carefully allows for the moving sediment/water interface. Then the effective sediment thickness, β^{-1}, is given by

$$\beta = [-w_s(1-f'+f'K_d)+\{w_s^2(1-f'+f'K_d)^2+4\lambda K_{eff}(1-f')(1-f'+f'K_d)\}^{\frac{1}{2}}]/(2K_{eff}(1-f'))$$

and

$$V_d = w_s(1-f'+f'K_d)+K_{eff}(1-f')\beta$$

The limit $\lambda=0$ yields the burial results given above, while $w_s = 0$ yields those for diffusive decay as long as one takes $(1-f')K_{eff} = K_{pw}$ and one notes that the more complete analysis given here replaces K_d by $1-f'+f'K_d$. The solution takes the burial or diffusive-decay limits as the parameter ϵ, defined by $\epsilon = w_s^2 (1-f'+f'K_d)/\lambda K_{eff} (1-f')$, is greater or less than one respectively. The case of burial combined with stirring equally of water and sediment as described by the diffusivity K_b in the bioturbated layer can similarly be treated by letting $(1-f')K_{eff} = (1-f'+f'K_d)K_b$.

97

EFFECTS OF BOUNDARY SCAVENGING PROCESSES

	Mixed Layer	Burial	Diff. / Decay
V_d	$h(1-f'+f'K_d)\lambda$	$w_s(1-f'+f'K_d)$	$\sqrt{K_d K_{pw}\lambda}$
Contrib. to λ_{eff}	$(1-f'+f'K_d)h\lambda/H$	$w_s(1-f'+f'K_d)/H$	$\sqrt{K_d K_{pw}\lambda}/\,H$
Eff. Sed. Thickness	$h(1-f'+f'K_d)/K_d$	$w_s(1-f'+f'K_d)/\lambda\,K_d$	$\sqrt{K_{pw}/(\lambda K_d)}$

formulae arises for short-lived substances which may not diffuse across the well-mixed layer before decaying. This is to say, their half-life is shorter than the mixing time for the layer. In this case, it is appropriate to use the diffusive-decay analysis given above but with K_{eff} replaced by K_b, the diffusivity of the bioturbated layer. Account of this effect can be taken by replacing h in the mixed-layer column by $(K_b/\lambda)^{\frac{1}{2}}$ whenever this is less than h. Burial and diffusion and decay in the underlying sediments may be neglected in this case.

Let us now estimate for a range of reactivities and radioactive half-lives which removal process should be most important. For this purpose the ratio is given for the removal by a process as compared to the removal by radioactive decay in the water column, assuming the bottom concentration is equal to the average water column concentration (that is that λ_{eff} as defined is appropriate). If this is not the case the ratio given in the Tables is not easily interpreted but the removal process giving the largest ratio will still be dominant.

The ratios are given in Tables A.V-IIa-c for each process for short- (30 a), medium- (1000 a), and long- (30 000 a) lived substances. The other parameters are taken to be h = 5 cm, H = 5 km, w_s = 1 cm ka^{-1}, f' = 1/2, K_v = 1 cm^2 s^{-1}, K_{pw} = 10^{-6} cm^2 s^{-1}, and K_b = 10^{-8} cm^2 s^{-1} throughout. When the effective length $(K_b/\lambda)^{\frac{1}{2}}$ is less than h, the diffusive limit is used for the well-mixed layer.

These results may in turn be summarized, as in Table A.V-III below, which shows the dominant processes and the ratio of the removal rate to decay in the water column when they are important.

From this Table it is apparent that for anything other than reactive substances, radioactive decay is the dominant removal process on an ocean-wide scale, unless the contaminant is stable (or very long-lived indeed) when net sediment accumulation (burial) is of course the only removal process. For reactive contaminants ($K_d \geqslant 10^6$) net sediment accumulation is generally the dominant process, although mixing into the surface layer may be dominant for non-retentive contaminants; (that is, those that are remobilized when calcareous or organic carrier particles are dissolved), particularly if they are also short-lived.

An interesting feature of this analysis is that diffusion into underlying sediments is never a dominant process over this range of parameter space. This may be understood by referring to Figure A.V-I.

TABLE A.V-IIa
RATIOS FOR REMOVAL PROCESSES
(Short-lived ($\lambda = 10^{-9}$ s^{-1}, $\lambda^{-1} = 30$ a))

Removal Process / Contaminant Type	Mixed Layer	Burial	Diff./ Decay
Soluble $K_d = 1$	6×10^{-6}	7×10^{-8}	6×10^{-5}
Moderate $K_d = 1000$	3×10^{-3}	3×10^{-5}	2×10^{-3}
Reactive $K_d = 10^6$	3	3×10^{-2}	6×10^{-2}
Highly Reactive $K_d \simeq 10^9$	3×10^3	30	2

TABLE A.V-IIb
RATIOS FOR REMOVAL PROCESSES
(Medium-lived ($\lambda = 3 \times 10^{-11}$ s^{-1}, $\lambda^{-1} = 1000$ a))

Removal Process / Contaminant Type	Mixed Layer	Burial	Diff./Decay
Soluble	10^{-5}	2×10^{-6}	4×10^{-4}
Moderate	5×10^{-3}	10^{-3}	10^{-2}
Reactive	5	1	0.4
Highly Reactive	5×10^3	10^3	10

TABLE A.V-IIc
RATIOS FOR REMOVAL PROCESSES
(Long-lived ($\lambda = 10^{-12}$ s^{-1}, λ^{-1} = 30 000 a))

Removal Process / Contaminant Type	Mixed Layer	Burial	Diff./Decay
Soluble	10^{-5}	7×10^{-5}	2×10^{-3}
Moderate	5×10^{-3}	3×10^{-2}	6×10^{-2}
Reactive	5	30	2
Highly Reactive	5×10^3	3×10^4	60

TABLE A.V-III
DOMINANT PROCESSES

Radioactive Lifetime / Contaminant Type	Short-Lived λ^{-1}= 30 a	Medium-Lived λ^{-1}= 1 000 a	Long-Lived λ^{-1} = 30 000 a	Stable
Soluble	Rad. Decay	Rad. Decay	Rad. Decay	Burial (∞)
Moderate	Rad. Decay	Rad. Decay	Rad. Decay	Burial (∞)
Reactive	Mixed Layer (3)	Mixed Layer or Burial (1-5)	Burial (30)	Burial (∞)
Highly Reactive	Mixed Layer (3×10^3)	Mixed Layer or Burial ($1-5 \times 10^3$)	Burial (3×10^4)	Burial (∞)

TABLE A.V-IV
DOMINANT DEPOSITION VELOCITIES
$(cm\ a^{-1})$

Radioactive Lifetime / Sediment Type	Short-Lived 30 a	Medium-Lived 1 000 a	Long-Lived 30 000 a	Stable
Soluble $K_d=1$	1 (Diff/ Decay)	0.2 (Diff/ Decay)	0.03 (Diff/ Decay)	0.001 (Burial)
Moderate $K_d=10^3$	100 (Mixed Layer)	2-5 (Diff/ Decay) (Mixed Layer)	0.5 - 1 (Diff/ Decay) (Burial)	0.5 (Burial)
Reactive $K_d=10^6$	10^5 (Mixed Layer)	2×10^3 (Mixed Layer)	5×10^2 (Burial)	5×10^5 (Burial)
Highly Reactive $K_d=10^9$	10^8 (Mixed Layer)	2×10^6 (Mixed Layer)	5×10^5 (Burial)	5×10^5 (Burial)

For short-lived substances the mixed-layer process is more effective, whilst for long-lived contaminants the net sediment accumulation process is more effective. There is, however, a region where diffusion and decay is the most important scavenging process as indicated by the deposition velocity, although it is less effective than radioactive decay in the water.

For future reference some typical values of deposition velocity have been calculated. As mentioned before, this is a useful parameter for interfacing the physical dispersion models to the sediments and is insensitive to the details of the interior distributions of the contaminant including the effects of particle scavenging. The flux obtained from the deposition velocity must of course be matched to the full flux by all mechanisms of the contaminant to the sediment water boundary.

The results presented in this Appendix, including the use of deposition velocities to parametrize removal mechanisms, have general applicability but one should note that they assume fast sediment mixing rates and chemical reaction times compared to other time-scales considered. In certain particular applications these assumptions will need to be treated with caution (see Appendix IX).

It is worth noting that when the basic assumptions regarding the deposition formalism are valid and for which the major sink of a substance lies on the bottom the relationship $V_dC(0) = Q/A$ must hold where $C(0)$ is the bottom concentration, A the bottom area, and Q the total source. Thus, given the proper formalism for V_d, the bottom concentration (or its average value) can be determined. It is also clear that in this case, no matter how badly the interior problem is solved, the bottom concentration will be correct and as such will, of course, be the total flux to the bottom Q/A.

In addition, it must be remembered that the flux into the sediments, as given by $V_d C(0)$, must be equal to the flux from the interior (unless this is modified by a surface source). At least for those cases where burial is an important sink relative to radioactive decay in the water column, particle scavenging must be included when modelling the interior concentration field. This is further pursued in Appendices VI, VII and IX.

Appendix VI

VERTICAL, ONE-DIMENSIONAL, ONE- AND TWO-LAYER BOUNDARY SCAVENGING MODELS

We envisage a sea-floor release, at a steady rate Q, of a substance with a finite decay rate λ. Without any scavenging the average concentration in the water must be $Q/\lambda V$, where V is the volume of the ocean. If the ocean mixing time is much less than λ^{-1} the substance concentration is actually uniform at a level $Q/\lambda V$, everywhere except very close to the source where much higher concentrations can be expected (see Appendices IV and VII).

If sedimentary scavenging, at the sea floor and in the water column, is now introduced we might expect that the concentration in the water is now reduced, perhaps as given by $Q/\lambda_{eff}V$, where λ_{eff}, as defined in Appendix V, allows for decay in the sediments as well as the water. In this Section, we examine the detailed effects of boundary scavenging on the results of one- and two-layer models, using the deposition velocity parametrization discussion in Appendix V. We also compare, in the context of a one-layer model, the effects of boundary and interior scavenging.

1. A ONE-LAYER MODEL WITH BOUNDARY SCAVENGING

A simple one-layer model with bottom sediment scavenging is obtained as a simple modification of Shepherd's illustrative 1-D model in IAEA-210. If K_V is the vertical diffusivity, z = H the surface, and z = 0 the bottom, the boundary condition at z = 0 is

$$- K_V \frac{dC}{dz} = \frac{Q}{A} - V_d C \qquad (A.VI.1)$$

where A is the area over which the source is distributed and V_d a deposition velocity as defined in Appendix V.

Solving the diffusion equation $K_V d^2C/dz^2 - C = 0$ with $dC/dz = 0$ at the surface z = H and Equation (A.VI.2) as the bottom boundary condition leads simply to the solution

$$C/C_o = (H/\Delta) \frac{\cosh{[(H-z)/\Delta]}}{\sinh{(H/\Delta)}} \left\{ 1 + (\frac{V_d}{\lambda H}) \frac{(H/\Delta)}{\tanh{(H/\Delta)}} \right\}^{-1} \qquad (A.VI.2)$$

with $\Delta = (K_V/\lambda)^{\frac{1}{2}}$ and $C_o = Q(V)^{-1}$. The term $\{\ \}$ represents the modification to the solution in IAEA-210 associated with boundary removal. It may be expressed alternatively as

$$\left\{ 1 + (\frac{K_d h^*}{H}) \frac{(H/\Delta)}{\tanh{(H/\Delta)}} \right\}^{-1} \qquad (A.VI.3)$$

103

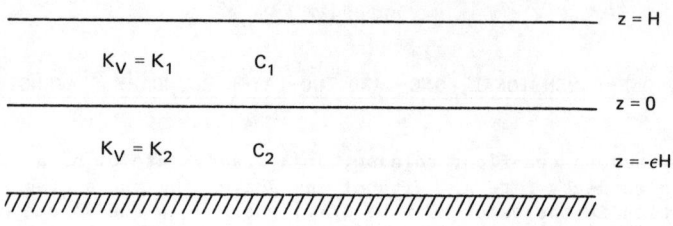

FIG. A.VI-1. *Boundary scavenging two-layer model.*

where

$$K_d h^* = V_d \ \lambda^{-1} \tag{A.VI.4}$$

represents the "equivalent water depth" of the scavenging sediment, and h^* is the effective sediment thickness of Appendix V. Equations (A.VI.2) and (A.VI.3) demonstrate that, in the 1-D case, scavenging at the sea floor does not alter the shape of the concentration profile, but reduces its magnitude. The reduction factor is just what one might have expected; for $H \ll \Delta$ the distribution in the water is uniform and the reduction factor is $(1+V_d(\lambda H)^{-1})^{-1} = \{1+K_d h^*/H\}^{-1}$, whereas for $H \gg \Delta$ the concentration in the water falls off over the scale depth and the reduction factor is $\{1+V_d(\lambda K_V)^{-\frac{1}{2}}\} = \{1+K_d h^*/\Delta\}^{-1}$. In each case the reduction factor is just as if one added to the effective water depth, H or Δ, the equivalent water depth $K_d h^*$ of the scavenging sediment.

2. A TWO-LAYER MODEL WITH BOUNDARY SCAVENGING

A better approximation to the real ocean would be a two-layer model, with the lower layer representing the benthic boundary layer with a much higher diffusivity than in the rest of the water column. We thus examine the model illustrated in Figure A.VI-1.

In each layer we must solve the equation

$$K_V \ d^2 C/dz^2 - \lambda C = 0 \tag{A.VI.5}$$

with boundary conditions

$$dC/dz = 0 \qquad \text{at} \qquad z = H \tag{A.VI.6}$$

$$- K_2 dC/dz = Q/A - V_d C \qquad \text{at} \qquad z = - \epsilon H \tag{A.VI.7}$$

C and $K_V dC/dz$ continuous at z = 0 $\hspace{2cm}$ (A.VI.8)

The solution is

$$C_1 = a_1 \cosh \left(\frac{H-z}{\Delta_1}\right), \quad \Delta_1 = (K_1/\lambda)^{\frac{1}{2}} \quad \text{(A.VI.9)}$$

$$C_2 = a_2 \cosh (-z/\Delta_2 + \phi), \Delta_2 = (K_2/\lambda)^{\frac{1}{2}} \quad \text{(A.VI.10)}$$

with Equation (A.VI.6) already satisfied.
The bottom boundary condition at $z = -\epsilon H$ implies

$$(K_2/\Delta_2)a_2 \sinh (\epsilon H/\Delta_2 + \phi) = Q/A - V_d a_2 \cosh (\epsilon H/\Delta_2 + \phi) \quad \text{(A.VI.11)}$$

or

$$a_2 = (Q/A) [(K_2/\Delta_2) \sinh (\epsilon H/\Delta_2 + \phi) + V_d \cosh (\epsilon H/\Delta_2 + \phi)]^{-1} \quad \text{(A.VI.12)}$$

The matching conditions at $z = 0$ imply

$$a_1 \cosh (H/\Delta_1) = a_2 \cosh \phi \quad \text{(A.VI.13)}$$

$$(K_1/\Delta_1) a_1 \sinh (H/\Delta_1) = (K_2/\Delta_2) a_2 \sinh \phi \quad \text{(A.VI.14)}$$

$$\therefore \tanh \phi = (K_1/K_2)^{\frac{1}{2}} \tanh (H/\Delta_1) \quad \text{(A.VI.15)}$$

We are particularly interested in the value of a_1, and its sensitivity to the presence of the bottom boundary layer.

$$a_1 = (Q/A) \cosh \phi \ \text{sech} (H/\Delta_1) [(K_2/\Delta_2) \sinh (\frac{\epsilon H}{\Delta_2} + \phi)$$

$$+ V_d \cosh (\frac{\epsilon H}{\Delta_2} + \phi)]^{-1} \quad \text{(A.VI.16)}$$

We first check the case with no bottom boundary layer ($\epsilon = 0$). This leads to

$$a_1 = (\frac{Q}{A\lambda H}) \frac{(H/\Delta_1)}{\sinh (H/\Delta_1)} \left[1 + \frac{V_d}{\lambda H} \frac{(H/\Delta_1)}{\tanh (H/\Delta_1)}\right]^{-1} \quad \text{(A.VI.17)}$$

as in Equation (A.VI.2). If we now assume $\epsilon H/\Delta_2 \ll 1$, that is that the mixing time over the bottom boundary layer is much less than the half-life of the substance, we have

$$a_1 = (Q/A) \ \text{sech} (H/\Delta_1) \left| (K_2/\lambda_2) [(\frac{\epsilon H}{\lambda_2} + \tanh \phi] + V_d [1 + \frac{\epsilon H}{\Delta_2} \tanh \phi]\right|^{-1}$$

$$= (\frac{Q}{A\lambda H}) \frac{(H/\Delta_1)}{\sinh (H/\Delta_1)} \left| 1 + (\frac{\epsilon H + K_d h^*}{H}) \frac{(H/\Delta_1)}{\tanh (H/\Delta_1)}\right|^{-1} \quad \text{(A.VI.18)}$$

where we have also assumed $\epsilon H^2/\Delta_2^2 \ll 1$ (i.e. H/Δ_2 not large) and define h* as in Equation (A.VI.3).

This solution shows that the only additional effect of a bottom boundary layer, with a mixing time short compared with the contaminant half-life, is to add its depth ϵH to the effective depth $K_d h^*$ of the scavenging sediment. The conclusions reached in Section 1 of this Appendix have therefore not changed significantly.

3. A ONE-DIMENSIONAL MODEL WITH INTERIOR AND BOUNDARY SCAVENGING

If the contaminant is also being scavenged in the ocean interior by particles falling with a velocity, w_p, and, if the contaminant on the particles is in equilibrium with that in the water, the governing equation in the ocean interior is

$$K_V^* \frac{d^2 C}{dz^2} - \lambda^* C + (V_i - w) \frac{dC}{dz} = 0 \qquad (A.VI.19)$$

where w is the upwelling velocity and

$$K_V^* = (1+\alpha)K_V$$

$$\lambda^* = (1+\alpha) \qquad (A.VI.20)$$

$$V_i = -\alpha w_p$$

wherein

$$\alpha = fK_d(1-f)^{-1} \qquad \text{(See also Appendix IX.)}$$

In Equation (A.VI.19) diffusion of particles has been included, but in reducing the expressions of Appendix IX, λ has been assumed to be small compared with k_2, the backward bulk rate constant. Note also that $f \ll 1$ and that usually $fK_d \ll 1$ even though scavenging is important.

The bottom boundary condition Equation (A.VI.1) now can be rewritten as

$$-K_V^* \frac{dC}{dz} - V_i C + wC = \frac{Q}{A} - V_d C + wC(H) \quad \text{at } z = 0 \qquad (A.VI.21)$$

where the additional terms on the left-hand side now include the flux of contaminant to the bottom by particles and away from the bottom by upwelling. The term wC(H) on the right-hand side allows for reinjection of water from the surface at the bottom (via deep upwelling). Equation (A.VI.21) is thus the equilibrium form of the bottom boundary condition of the model of Appendix IX in the limit that the thickness of the bottom boundary layer is very small.

In the absence of a surface source, the surface boundary condition is similarly given by

$$K_V^* \frac{dC}{dz} + V_i C = 0 \qquad \text{at } z = H \qquad\qquad \text{(A.VI.22)}$$

The solutions of (A.VI.19) are given by

$$C = a\, e^{r_1 z} + b\, e^{r_2 z} \qquad\qquad \text{(A.VI.23)}$$

where r_1 and r_2 are the roots of

$$K_V^* r^2 - w^* r - \lambda^* = 0 \qquad \text{wherein } w^* = w - V_i. \qquad \text{(A.VI.24)}$$

They are given by

$$r_{1,2} = \frac{w^*}{K_V^*} \left[\frac{1}{2} \pm \frac{1}{2} (1 + \frac{4 K_V^* \lambda^*}{w^{*2}})^{\frac{1}{2}} \right] \qquad\qquad \text{(A.VI.25)}$$

The surface boundary condition yields

$$a\,(K_V^* r_1 + V_i)\, e^{r_1 H} + b\,(K_V^* r_2 + V_i)\, e^{r_2 H} = 0 \qquad\qquad \text{(A.VI.26)}$$

and the bottom boundary condition yields

$$- [K_V^* r_1 - (w^* + V_d)]\, a - [K_V^* r_2 - (w^* + V_d)]\, b \qquad\qquad \text{(A.VI.27)}$$

$$= \frac{Q}{A} + (w^* + V_i)(a\, e^{r_1 H} + b\, e^{r_2 H})$$

Combining Equations (A.VI.26) and (A.VI.27) gives

$$b = \frac{Q}{A} \left\{ (w^* + V_i) \left[\left(\frac{K_V^* r_2 + V_i}{K_V^* r_1 + V_i} \right) - 1 \right] e^{r_2 H} \right.$$

$$\text{(A.VI.28)}$$

$$\left. + (K^* r_1 - w^* - V_d) \left(\frac{K_V^* r_2 + V_i}{K_V^* r_1 + V_i} \right) e^{(r_2 - r_1)H} - (K_V^* r_2 - w^* - V_d) \right\}^{-1}$$

A similar (symmetrical in r_1 and r_2) expression for a may be obtained from Equation (A.VI.26) and the concentration field evaluated from Equation (A.VI.23). This provides the most general one-dimensional equilibrium model presented in this report. Its overall validity may be understood using the analysis of Appendix IX.

107

The relative role of boundary and interior scavenging may be examined by considering the situation where there is no upwelling ($w=0$) and where the contaminant is confined within a scale height, r^{-1}, much less than the depth of the ocean. This will be the case if H is greater than either the vertical scale, $K_V^* V_i^{-1}$, associated with a balance between scavenging and diffusion or the scale, $(K_V/\lambda^*)^{\frac{1}{2}}$, associated with decay and diffusion. The solution is then given by

$$C = \frac{Q}{A} (V_d - V_i + K_V^* r)^{-1} e^{-rz}$$

(A.VI.29)

where

$$r = \frac{V_i}{K_V^*} [\frac{1}{2} + \frac{1}{2} (1 + \frac{4K_V^* \lambda^*}{V_i^2})^{\frac{1}{2}}]$$

Let us first consider the relationship between scavenging in the interior, as parametrized by V_i, and burial by the sediments as part of V_d. If the contributions to V_d could be considered as additive, one could define $V_b = V_d - V_d^{burial}$, where V_b includes the effects of diffusion and mixing into the bioturbated layer. Consider now the case that $K_d \gg 1$ and $f \ll 1$. Then $V_d^{burial} = w_s f' K_d$, where K_d' is the sedimentary K_d, and $V_i = -w_p f K_d$. One may also write $-w_s f' = \delta w_p f$ where $\delta \leqslant 1$ includes the effect of particulate dissolution (say of the organic fraction). This may or may not lead to the release of the contaminant to the pore waters.

Since our general understanding is that burial of a reactive substance will not exceed its rate of arrival to the sediments on particles, the ratio $(V_d^{burial}/V_i) = (\delta K_d'/K_d) = \epsilon \leqslant 1$. If the sediments are retentive $\epsilon \simeq 1$ and $V_d^{burial} \simeq V_i \simeq w_s f' K_d / \delta$.

The factor $(V_d - V_i + K_V^* r)^{-1}$ in Equation (A.VI.29) that determines the bottom value of C may now be rewritten as $\left(V_b + V_i [\epsilon - \frac{1}{2} + \frac{1}{2}(1 + \frac{4\lambda^* K_V^*}{V_i^2})^{\frac{1}{2}}]\right)^{-1}$.

This clearly indicates the dependence of the bottom concentration, especially for long-lived reactive substances, on ϵ and thus on the fate of particulate matter and the contaminant it carries when it enters the bottom boundary layer. On the other hand the vertical scale height in the water column, r^{-1}, only depends on V_i, which describes scavenging in the ocean interior and is independent of boundary processes.

Appendix VII

THE EFFECTS OF STRONG LOCAL SCAVENGING

As discussed in Section 2.3 and in Appendix IV, there is inevitably a region of substantially elevated concentration in the vicinity of a localized ("point") source. Scavenging processes (of any order greater than zero) would lead to correspondingly elevated fluxes of contaminant being removed to the bottom in the vicinity of the source. It is at first sight possible that sufficient contaminant might be removed that very little is able to disperse into the remainder of the ocean, and that the localization of the scavenging may serve to enhance this effect.

It has been shown in Appendix VI, using one-dimensional models, that it is indeed possible for scavenging to be so strong that the greater part of a contaminant is removed to the sediments. Such models cannot however tell us whether the localized elevations of the concentrations are of any consequence, since they do not allow any representation of them.

In this Appendix we demonstrate that the localized region of high concentration has no effect on the partitioning of contaminant between water and sediments - the reduction of ocean inventory due to scavenging is just what one would estimate from the one-dimensional models. The extent to which the inventory of contaminant residing on the sediments in the immediate vicinity may nevertheless be a large fraction of the total, is also discussed, using some results of a model of Garrett and Shepherd (1986).

1. POSSIBLE MODIFICATION OF OCEAN INVENTORY

Consider an ocean in which contaminant is moving around, described by a flux field $\underline{F}$, due to advection, diffusion and sedimentary processes. Everywhere in the interior of the ocean (i.e. away from any sources), the divergence of the flux field must be balanced by radioactive decay, since contaminant is otherwise conserved. Therefore:

$$\text{div } \underline{F} - \lambda C = 0 \qquad\qquad (A.VII.1)$$

Separating horizontal and vertical components, one has:

$$\nabla_H \cdot \underline{F}_H + \frac{\partial F_V}{\partial z} - \lambda C = 0 \qquad\qquad (A.VII.2)$$

where the suffixes H and V here indicate the horizontal and vertical components.

Integrating over any horizontal surface of area A spanning the entire ocean:

$$\int_A \nabla_H \cdot \underline{F}_H \, da + A \frac{d\overline{F}_V}{dz} - \lambda \overline{C} = 0 \qquad\qquad (A.VII.3)$$

109

where $^{-}$ indicates the horizontal average. However, by Green's theorem:

$$\int_A \nabla_H \cdot \underline{F}_H \; da \;\; = \;\; \int \underline{F}_H \cdot \underline{n} \; dl \qquad\qquad\qquad \text{(A.VII.4)}$$

which is the integral around the perimeter of the horizontal flux normal to the boundary. Since (in the absence of deposition on the sides) this normal flux of contaminant is zero everywhere (no transport in or out of the sides), the right-hand side is identically zero.

The interior processes are therefore correctly summarized by the vertical flux balance equation for the horizontally averaged concentration:

$$\frac{d\overline{F}_V}{dz} \;\; - \;\; \lambda \, \overline{C} \;\; = \;\; 0 \qquad\qquad\qquad \text{(A.VII.5)}$$

which is exactly the equation treated by the one-dimensional models of Appendix VI, provided that the vertical diffusivity is horizontally uniform.

Similarly, at the bottom boundary, the total depositional flux is (using the deposition velocity parametrization) in the absence of sources, just:

$$\int_A F_V(0) da \;\; = \;\; \int_A V_d \; C(0) da \qquad\qquad\qquad \text{(A.VII.6)}$$

Taking V_d constant over the surface (which does not affect the essence of the problem), this reduces to:

$$\overline{F}_V(0) \;\; = \;\; V_d \; \overline{C}(0) \qquad\qquad\qquad \text{(A.VII.7)}$$

Thus the bottom boundary condition can also be written entirely in terms of vertical fluxes of horizontally averaged quantities.

These results mean that the oceanic inventory of a contaminant (but not of course its horizontal distribution) can be deduced from a one-dimensional treatment of average quantities, so that the results for the reduction of ocean inventory due to scavenging derived in Appendix VI are in fact valid more generally than might have been expected.

Since the actual horizontal distributions of concentrations are of no consequence (provided the average remains the same) this means that the enhancement of scavenging over and above that which would be estimated from a one-dimensional model cannot in fact occur.

This conclusion remains true whether the interior transport is primarily advective, diffusive or particulate, and provided the removal processes are zeroth- or first-order, even if they are reversible. It would fail however if the removal process were of fractional order, or of order greater than one, or if there is substantial horizontal variability of the deposition velocity, vertical diffusivity or upwelling speed.

110

2. A SIMPLE THREE-DIMENSIONAL MODEL WITH BOUNDARY SCAVENGING

Neglecting interior scavenging for the moment, the concentration field C in a purely diffusive ocean satisfies the equation

$$K_H \nabla_H^2 C + K_V \partial^2 C / \partial z^2 - \lambda C = 0 \qquad \text{(A.VII.8)}$$

The boundary condition on the flat sea floor at z=0 is

$$- K_V \partial C / \partial z = Qf(r) - V_d C \qquad \text{(A.VII.9)}$$

for a distributed source that is assumed to be radially symmetric about the origin and of total strength Q if $\int_0^\infty 2\pi r f(r)dr = 1$.

For a cylindrical ocean of radius R, with $\partial C / \partial r = 0$ on r = R, a Fourier-Bessel solution of Equation (A.VII.8) may be obtained (Garrett and Shepherd, 1986). For zero (or small) values of V_d the solution resembles that described in Appendix IV: a local source-like behaviour superimposed on a larger-scale concentration field which is uniform over the ocean for long-lived contaminants, and in general has a horizontal average equal to that obtained from a one-dimensional model. The new question is whether a finite V_d can remove a contaminant before it diffuses out to the sides and top of the ocean, and so greatly modify the contaminant concentration in the far field from what it would be without scavenging. This may be investigated by examining the solution of Equations (A.VII.8) and (A.VII.9) for an infinite ocean. For a top-hat source defined by f(r)=1 for $r < r_s$ and f(r)=0 for $r > r_s$, the solution is

$$C = Q(\pi r_s)^{-1}(K_H K_V)^{-\frac{1}{2}} C',$$

$$C' = \int_0^\infty [\sigma + (k^2 + k_0^2)^{\frac{1}{2}}]^{-1} J_1(k) J_0(kr') \exp[-(k^2 + k_0^2)^{\frac{1}{2}} z'] dk \qquad \text{(A.VII.10)}$$

where $k_0 = (\lambda r_s^2 / K_H)^{\frac{1}{2}}$, $\sigma = V_d r_s (K_H K_V)^{-\frac{1}{2}}$, $r' = r/r_s$, $z' = (K_H / K_V)^{\frac{1}{2}} (z/r_s)$

The normalizing factor $Q(\pi r_s)^{-1} (K_H K_V)^{\frac{1}{2}}$ in Equation (A.VII.10) is just the point-source solution from Appendix IV evaluated at a horizontal distance $\frac{1}{2} r_s$ from the origin; it expresses the averaging effect of a finite source region. In fact, for a source region small compared to the lateral diffusion/decay scale ($k_0 \ll 1$) as it will usually be, and for $\sigma \ll 1$, as it will be for deposition velocities less than about 30 m a^{-1} given r_s = 100 km, $K_H = 10^2$ m^2 s^{-1} and $K_V = 10^{-4}$ m^2 s^{-1}, we find that $C' \approx 1$ close to the source and that the singularity in Appendix IV may be replaced by the value of the point-source solution at $r = \frac{1}{2} r_s$.

The formula (A.VII.10) may be evaluated numerically, but the results are best comprehended through various approximate formulae for C' which are valid in the regions $1 \ll \rho \ll \sigma^{-1}$ and $\sigma^{-1} \ll \rho$, with $\rho = (r'^2 + z'^2)^{\frac{1}{2}}$. These results, summarized in Figure A.VII-1, are derived by Garrett (1983) and have been checked against numerical integration of Equation (A.VII.10). The existence of the intermediate

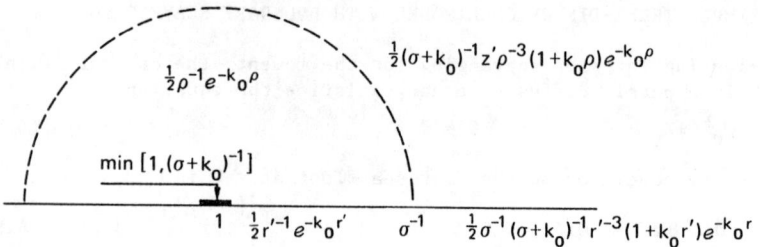

FIG. A.VII-1. *Approximate formulae for the normalized concentration from the solution (A.VII.10) of the boundary scavenging problem, in the regions close to the origin, $1 \ll \rho \ll \sigma^{-1}$ and $\sigma^{-1} \ll \rho$.*

region $1 \ll \rho \ll \sigma^{-1}$ depends, of course, on having $\sigma \ll 1$. In this region, if it exists, the details of the source are unimportant and scavenging has not had time to affect the concentration field significantly. Hence the concentration is identical with that for a point source of a decaying contaminant, as discussed in Appendix IV.

For $\rho \gg \sigma^{-1}$, (or $\rho \gg 1$ if $\sigma \gg 1$), the concentration field away from the bottom is that of a dipole (with extra terms to allow for decay), as would be given by a point source above a perfectly absorbing boundary. If applied at the bottom ($z' = 0$) the interior formula would give $C' = 0$; this however is invalid and Figure A.VII-1 shows the next non-zero approximation for $C'(r',0)$ in the far field. In the vicinity of the origin finite values of σ or k_0 reduce C' from the value of 1 which it takes for $\sigma = k_0 = 0$. For $\sigma \gg 1$ or $k_0 \gg 1$, $c' \sim \sigma^{-1}$ or k_0^{-1}, so that a reasonable approximation, as shown in Figure A.VII-1, is $C'(0,0) \approx \min[1, (\sigma+k_0)^{-1}]$.

The important result of this model is the identification of the scale σ^{-1} beyond which the concentration falls off rapidly. In dimensional terms this is $K_V V_d^{-1}$ vertically and $(K_H K_V)^{\frac{1}{2}} V_d^{-1}$ horizontally. Of course the discussion in Section 1 of this Appendix shows that the horizontally averaged concentration field in a finite ocean is given by the one-dimensional solution of Appendix VI, so the main advance here is the identification of the key horizontal scale $(K_H K_V)^{\frac{1}{2}} V_d^{-1}$ beyond which the effects of scavenging on the distribution of concentration become apparent.

3. THE EFFECT OF INTERIOR SCAVENGING

If interior scavenging by sinking particles is included and the only removal at the bottom is by burial of the sinking sediments, the governing equations are (assuming f and $fK_d \ll 1$)

$$K_H \nabla_H^2 C + K_V \partial^2 C / \partial z^2 - \lambda C + V_i \partial C / \partial z = 0 \qquad \text{(A.VII.11)}$$

with boundary condition

$$-K_V \partial C / \partial z = Qf(r) \qquad \text{(A.VII.12)}$$

as for Equation (A.VI.21) with $V_d = V_i$.

112

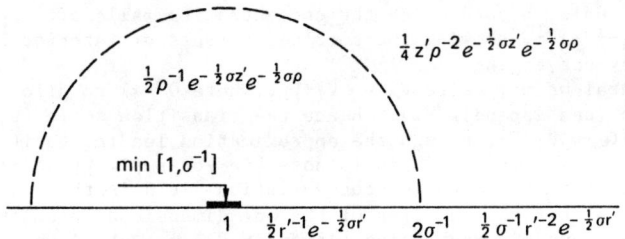

$$\tfrac{1}{4}z'\rho^{-2}e^{-\frac{1}{2}\sigma z'}e^{-\frac{1}{2}\sigma\rho}$$

$$\tfrac{1}{2}\rho^{-1}e^{-\frac{1}{2}\sigma z'}e^{-\frac{1}{2}\sigma\rho}$$

min $[1,\sigma^{-1}]$

1    $\tfrac{1}{2}r'^{-1}e^{-\frac{1}{2}\sigma r'}$ $2\sigma^{-1}$ $\tfrac{1}{2}\sigma^{-1}r'^{-2}e^{-\frac{1}{2}\sigma r'}$

FIG. A.VII–2. *Approximate formulae for the normalized concentration from the solution (A.VII.13) of the interior scavenging problem with $k_0 = 0$, in the regions close to the origin, $1 \ll \rho \ll 2\sigma^{-1}$ and $2\sigma^{-1} \ll \rho$.*

This problem may also be solved for a finite ocean using a Fourier–Bessel expansion; here we concentrate on the solution for an infinite ocean. This is (Garrett and Shepherd, 1986)

$$C = Q(\pi r_s)^{-1}(K_H K_V)^{-\frac{1}{2}}C'$$

$$C' = \int_0^{\infty} [\,\tfrac{\sigma}{2} + (\tfrac{\sigma^2}{4} + k^2 + k_0^2)^{\frac{1}{2}}\,]^{-1} J_1(k) J_0(kr') \exp\left|-[\tfrac{\sigma}{2} + (\tfrac{\sigma^2}{4} + k^2 + k_0^2)^{\frac{1}{2}}]z'\right| dk \quad \text{(A.VII.13)}$$

where k_0, r', z', are as before, and $\sigma = V_i r_s (K_H K_V)^{-\frac{1}{2}}$. We see that C' in this case is just $\exp(-\tfrac{1}{2}\sigma z')$ times the solution (A.VII.10) for boundary scavenging, with σ in (A.VII.10) replaced by $\tfrac{1}{2}\sigma$, and k_0 replaced by

$$\left(\tfrac{\sigma^2}{4} + k_0^2\right)^{\frac{1}{2}}$$

Hence the formulae in Figure A.VII–1 may be adapted to the problem for interior scavenging. In Figure A.VII–2 we show the results for the case $k_0 = 0$ so that the decay rate of the contaminant is zero and the only removal process is at the bottom. The transition scale is much as when there is no interior scavenging, but with the extra decay terms $e^{-\frac{1}{2}\sigma z'}$ and $e^{-\frac{1}{2}\sigma\rho}$. Thus, interior scavenging enhances removal near the source. Since $V_d = V_i$ in this problem, the key result is again the identification of scales of order $(K_H K_V)^{\frac{1}{2}}/V_d$ horizontally and K_V/V_d vertically within which most of the contaminant is contained.

If we allow for interior scavenging and general boundary removal processes (that is $V_i \neq V_d$), we must solve Equation (A.VII.11) with boundary condition

$$-V_i C - K_V \frac{\partial C}{\partial z} = Qf(r) - V_d C$$

similar to Equation (A.VI.21). A closed form for the solution, similar to Equations (A.VII.10) and (A.VII.13), can be written down and its asymptotic behaviour in different regions investigated. The horizontal transition scale, beyond which the concentration falls off rapidly, is $(K_H K_V)^{\frac{1}{2}} (V_d - \frac{1}{2} V_i)^{-1}$, showing the combined effects of interior and boundary scavenging.

In general we may write $V_d = \epsilon V_i + V_b$, where $0 < \epsilon \leqslant 1$ to allow for dissolution (see Appendix VI). Hence the transition scale is $(K_H K_V)^{\frac{1}{2}} [V_b + (\epsilon - \frac{1}{2}) V_i]^{-1}$, though the approximation leading to it breaks down if $V_b + (\epsilon - \frac{1}{2}) V_i < 0$, as is possible for small V_b and ϵ. Suitable approximations to the concentration field in this case are under investigation, but the solution of the one-dimensional problem with interior and boundary scavenging (Appendix VI) can still be used for the laterally averaged concentration field.

However, as long as $V_b + (\epsilon - \frac{1}{2}) V_i > 0$, the solutions are similar in form to $\exp(-\frac{1}{2} \sigma_i z')$ times the integral in A.VII.10 with σ replaced

by σ_* (equal to $\sigma_b + \frac{1}{2} \sigma_i$), and k_o replaced by k_* (equal to $(k_o^2 + \frac{\sigma_i^2}{4})^{\frac{1}{2}}$),

where $\sigma_i = V_i r_s (K_H K_V)^{-\frac{1}{2}}$, $V_b = V_d - V_i$, and $\sigma_b = V_b r_s (K_H K_V)^{-\frac{1}{2}}$.

Noting that there is no essential difference between the solutions on and off the bottom, except that for $\rho > 1/\sigma_*$ the term z' reduces to a lower limit of $1/\sigma_*$ as the bottom is approached, the asymptotic solutions may be written in three regions, i.e.

$$C' = (1 + \sigma_* + k_*)^{-1} ; \qquad \rho < 1$$

$$C' = \frac{1}{2} \exp(-\frac{1}{2} \sigma_i z') \, \rho^{-1} \, \exp(-k_* \rho); \ 1 < \rho < \sigma_*^{-1}$$

$$C' = \frac{1}{2} \exp(-\frac{1}{2} \sigma_i z')(\sigma_* + k_*)^{-1}(\sigma_*^{-1} + z') \rho^{-3}(1 + k_* \rho)\exp(-k_* \rho); \rho > \sigma_*^{-1} \qquad \text{(A.VII.14)}$$

These analytic solutions may be evaluated without difficulty as required (remembering that $C = Q(\pi r_s)^{-1} (K_H K_V)^{-\frac{1}{2}} C'$), and constitute the most general solution for near-field concentrations obtained by this method. The horizontal transition scale is r_s/σ_*.

4. SUMMARY

We have shown in Appendix VI that most of the contaminant will end up on the sediments if $V_d \gg \lambda H$ or $V_d \gg (K_V \lambda)^{\frac{1}{2}}$ for long- and short-lived contaminants, respectively ($K_V >$ or $< \lambda H^2$). In this Appendix it has been shown that the contaminant is retained by scavenging nearer to the source if the scavenging distance $(K_H K_V)^{\frac{1}{2}} V_d^{-1} \ll R$, the ocean radius, (or equivalently $K_V V_d^{-1} \ll H$ when $(K_V R^2)(K_H H^2)^{-1}$ is of order one). For short-lived contaminants the comparison is really between the scavenging scale (K_V/V_d) and the decay scale $(K_V/\lambda)^{\frac{1}{2}}$. The scavenging scale is less when the sedimentary inventory dominates $(V_d \gg (K_V \lambda)^{\frac{1}{2}})$. The situation is summarized in Figure A.VII-3 in which the ocean is categorized by the two dimensionless parameters $V_d H/K_V$ and $(\lambda H^2/K_V)^{\frac{1}{2}}$ representing separately the strength of scavenging processes and decay relative to vertical mixing.

In region A of Figure A.VII-3 the contaminant is mostly removed onto local sediments (and following the discussion in Appendix V this is likely to be due to burial for small values of $(\lambda H^2/K_V)^{\frac{1}{2}}$ and sorption by the bioturbated layer for large values). In region B the sedimentary

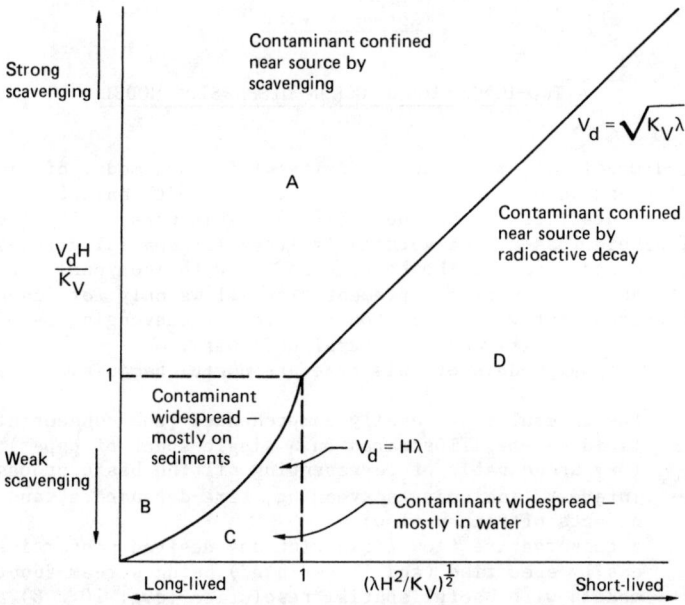

FIG. A. VII-3. Regions of parameter space with different dominant mechanisms of contaminant removal. The contaminant is largely associated with: local sediments in region A, global sediments in region B, the whole ocean in region C and the local ocean in region D.

inventory dominates the water inventory, but the contaminant is spread ocean-wide. In region C the contaminant is spread ocean-wide, but with the water inventory dominating the sedimentary inventory. The water inventory also dominates in region D, but the contaminant decays before it can spread ocean-wide. The transition from one region to another is, of course, more gradual than suggested by Figure A.VII-3, but the identification of the different types of behaviour in different regions of parameter space can be of considerable use.

Appendix VIII

A TWO-DIMENSIONAL OCEAN DISPERSION MODEL

A two-dimensional (meridional) finite-difference model of dispersion by diffusion and advection in an idealized ocean with an arbitrary steady flow has been described by Shepherd (1983). This uses a simple explicit numerical scheme, based on a vertically irregular spatial grid, so that levels in the model can be chosen to coincide with isopycnal rather than level surfaces. The current implementation allows only for scavenging at the ocean bottom, but generalization to interior scavenging by one or more particulate species would be straightforward.

Two-dimensional models of this type are useful because:

(a)   their results are easily comprehended (the concentration field may be illustrated on a single sheet of paper);

(b)   they are capable of representing all the basic processes of interest, including scavenging, time-dependence, and the effects of return flow;

(c)   a conservative flow field with any desired properties is easily specified (and illustrated) using stream functions;

(d)   models with useful spatial resolution (e.g. 10 x 8) are small and fast enough to enable large numbers of cases to be studied easily and cheaply.

The principal disadvantages are that such models inherently assume rapid mixing in the dimension not modelled (usually zonal), and subsume various elements of the oceanic circulation, even of quite large scale (e.g. anticyclonic gyres) into elevated mixing coefficients. The results must therefore be regarded as primarily illustrative, and of quantitative value only under closely specified conditions.

Among the results presented by Shepherd (1983) are some illustrating the effect of increasing boundary scavenging in a static diffusive ocean. One set of these results is reproduced here in Figure A.VIII-1, which shows the concentration fields obtained for a contaminant with a half-life of 300 a in an ocean with both vertical and horizontal mixing times of around 3000 a, for deposition velocities of zero, 1000 and 1×10^6 cm a^{-1}. Because the source is effectively somewhat above the bottom in these calculations, the scavenging may be limited by diffusion to the bottom, and the effective deposition velocities are roughly zero, 500 and 1000 cm a^{-1}. The figures do however clearly show the progressive modification of the field, and clearly illustrate some of the features derived analytically in Appendix VII.

116

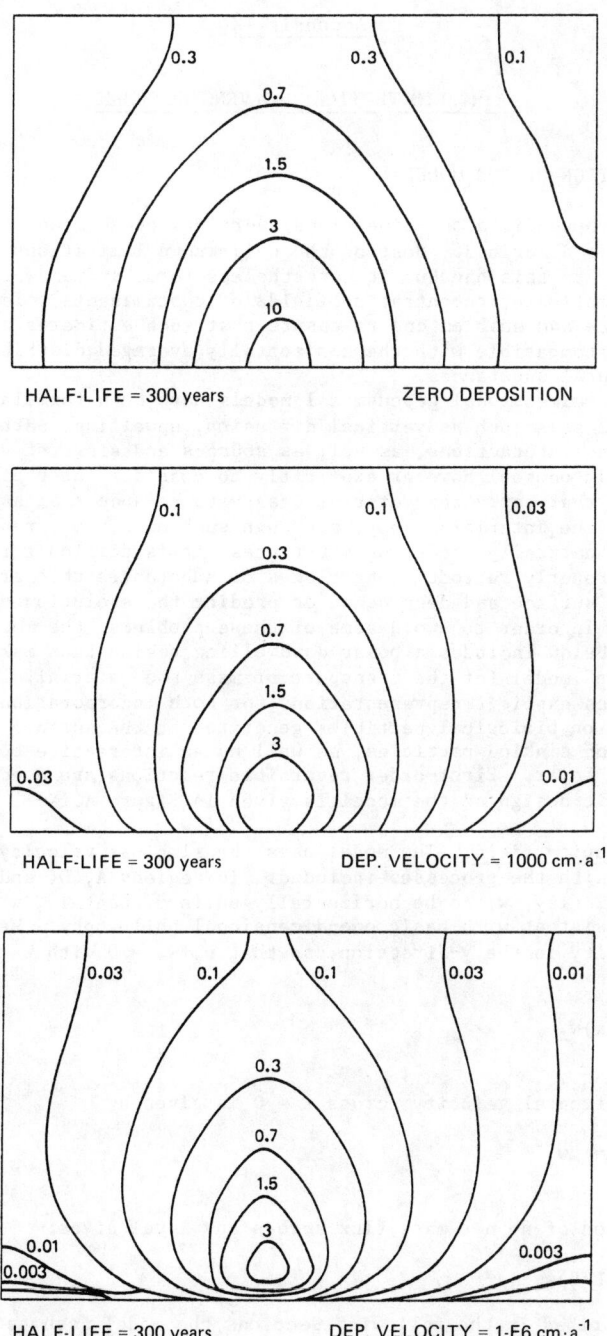

HALF-LIFE = 300 years   ZERO DEPOSITION

HALF-LIFE = 300 years   DEP. VELOCITY = 1000 cm·a⁻¹

HALF-LIFE = 300 years   DEP. VELOCITY = 1-E6 cm·a⁻¹

FIG. A.VIII–1.   Modification of concentration fields, due to depositional velocity.

Appendix IX

A HYBRID VERTICAL SCAVENGING MODEL

1. FORMULATION OF THE MODEL

In this Appendix a one-dimensional vertical ocean model including
scavenging is described. Most of the well-known limitations of such
models apply to this one but it nevertheless forms a framework that can
be used to estimate concentration fields of contaminants released from a
bottom source and enables one to ensure that such estimates are to the
first order compatible with the horizontally averaged distributions of
various natural substances.

Many one-dimensional geochemical models have been formulated that
include processes such as vertical diffusion, upwelling, particulate
transport and interactions, as well as sources and sinks of various
types. It is unusual however explicitly to consider the regions of
downwelling that carry the water necessary to balance that assumed to be
upwelled in the interior. Thus, although such models may reproduce the
profiles of naturally occurring substances, it is difficult to ensure
that they properly reproduce the fluxes of substances that are recycled
between the surface and deep ocean or predict the evolution of a
transient. In order to avoid some of these problems, the model
formulated below includes a polar downwelling region thus making a
"closed-loop" model for the transport of mass and material. The model
also includes explicit representations for both incorporation of a
contaminant on biological particles generated at the surface and its
adsorption on sinking particles, as well as an interactive bottom
sedimentary layer. First-order reversible reactions are used throughout.

The basic design of the model is given in Figure A.IX-1.

The velocity field. The model uses the simplest velocity field
consistent with the processes included. In regions A, D, and E we take
vertical velocity, w, to be horizontally uniform, that is, $w = w(z)$.
This is consistent with basic one-dimensional philosophy. We also ignore
non-uniformity in the y-direction, so that $u_x + w_z = 0$ with $u = 0$ at
$x = L$. Thus

$$u = (L-x)w_z \tag{A.IX.1}$$

and the horizontal velocity across $x = 0$ is given by

$$u(x=0) = Lw_z \tag{A.IX.2}$$

The condition of no net mass flux across any level gives:

$$w' = -(L/\ell)w \tag{A.IX.3}$$

As described in the following Sections the model assumes in addition
that there is no entrainment from the interior into Region C. This
allows certain simplifications to be made (since then $w_z = 0$ for
$0 > z > -H + z_1$). In the region $-H + z_1 > z > -H$, $w(z)$ must be

118

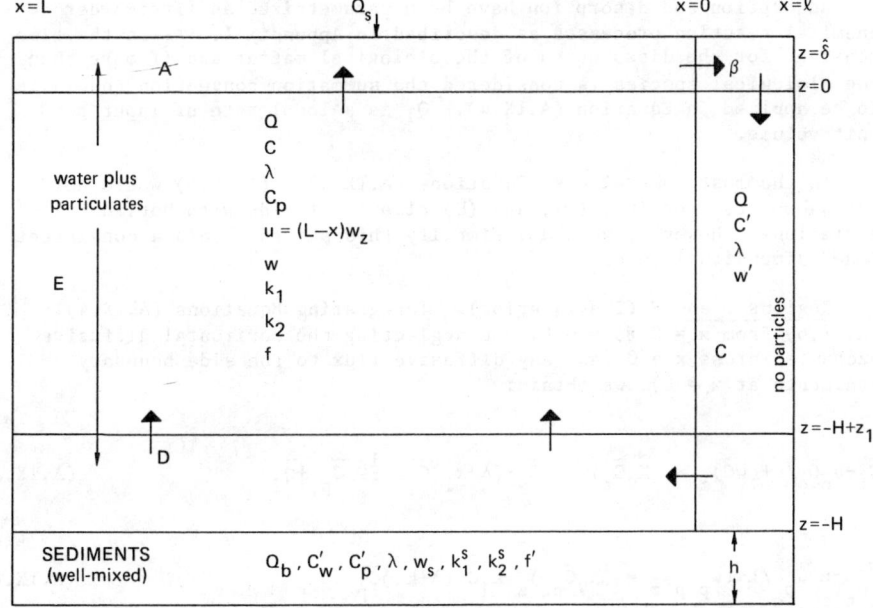

FIG. A.IX-1. One-dimensional vertical ocean model with scavenging.

specified. The simplest way to do this is to allow w to vary linearly from the necessary zero on the bottom to the interior value at $z = -H + z_1$.

The basic equations. The substance under consideration is assumed to exist in three different states: (1) dissolved in the water; (2) adsorbed onto the exterior of particles in the water column; and (3) incorporated into biological matter which rains out of the surface layer, dissolving as it falls to the bottom. All the organic matter which reaches the bottom is then assumed to dissolve there. The associated equations are:

$$C_t + (uC)_x + (wC)_z = (K_H C_x)_x + (K_V C_z)_z + k_2 C_p - (\lambda + k_1)C + \frac{1}{\tau_i} C_{p_i} + Q_I \qquad (A.IX.4)$$

$$C_{p_t} + (uC_p)_x + (w_p C_p)_z = (K_H C_{px})_x + (K_V C_{p_z})_z + k_1 C - (\lambda + k_2)C_p \qquad (A.IX.5)$$

$$C_{p_{i_t}} + (w_{p_i} C_{p_i})_z = -\lambda C_{p_i} - \frac{1}{\tau_i} C_{p_i} \qquad (i=1,N) \qquad (A.IX.6)$$

where the usual symbols have been used with the addition that w_{p_i} is the settling velocity of the biological matter and C_{p_i} its concentration.

119

Adsorption and desorption have been parametrized as first-order chemical reaction processes as described in Appendix I. τ_i is the time constant for the dissolution of the biological matter and if more than one biological species is considered the summation convention (on i) is to be applied in Equation (A.IX.4). Q_I is a local rate of input per unit volume.

In the most general case Equations (A.IX.4) - (A.IX.6) would hold throughout regions (C), (D), and (E) allowing fields with horizontal variations. However, we shall simplify this set to obtain a consistent "one"-dimensional model.

Regions D and E (The Interior). Integrating Equations (A.IX.4) - (A.IX.6) from x = 0 to x = L, and neglecting the horizontal diffusive exchange across x = 0 (and any diffusive flux to the side boundary sediments at x = L), we obtain:

$$\bar{C}_t - u_o C_o / L + (w\bar{C})_z = (K_v \bar{C}_z)_z + k_2 \bar{C}_p - (\lambda + k_1)\bar{C} + \frac{1}{\tau_i}\bar{C}_{p_i} + \bar{Q}_I \tag{A.IX.7}$$

$$\bar{C}_{p_t} - u_o C_{p_o} / L + (w_p \bar{C}_p)_z = (K_v \bar{C}_{pz})_z + k_1 \bar{C} - (\lambda + k_2)\bar{C}_p \tag{A.IX.8}$$

$$\bar{C}_{p_{i_t}} + (w_{p_i} \bar{C}_{p_i})_z = -(\frac{1 + \lambda \tau_i}{\tau_i})\bar{C}_{p_i} \tag{A.IX.9}$$

where the subscript "o" indicates values at x = 0 and "⁻" the horizontal average.

Neglecting the influence of entrainment from Region E into the northern boundary layer, i.e. taking $w_z = u_o/L = 0$ for $-H + z_1 < z < 0$, and taking w_p and w_{p_i} to be constant, yields

$$\bar{C}_t + w\bar{C}_z = (K_v \bar{C}_z)_z + k_2 \bar{C}_p - (\lambda + k_1)\bar{C} + (1/\tau_i)\bar{C}_{p_i} + \bar{Q}_I \tag{A.IX.10}$$

$$\bar{C}_{p_t} + w_p \bar{C}_{pz} = (K_v \bar{C}_{pz})_z + k_1 \bar{C} - (\lambda + k_2)\bar{C}_p \tag{A.IX.11}$$

$$\bar{C}_{p_{i_t}} + w_{p_i} \bar{C}_{p_{iz}} = (\frac{1 + \lambda \tau_i}{\tau_i})\bar{C}_{p_i} \tag{A.IX.12}$$

in Region E.

In Region D, Equations (A.IX.7) - (A.IX.9) still hold but this region is supplied at x = 0 by water from the polar downwelling region. Thus we take $C_0 = C'$, $C_{p_0} = 0$ (see Regions C and F) and obtain from Equation (A.IX.7)

$$\overline{C}_t - w_z \overline{C}' + (w\overline{C})_z = (K_V \overline{C}_z)_z + k_2 \overline{C}_p - (\lambda + k_1)\overline{C} + (1/\tau_i)\overline{C}_{p_i} + \overline{Q}_I \qquad \text{(A.IX.13)}$$

in Region D.

Equations (A.IX.8) and (A.IX.9) reduce to the same forms, Equations (A.IX.11) and (A.IX.12) respectively, in D as in E.

(Note that $w_{pz} = w_{p_{iz}} = 0$ but w_z must be $\neq 0$ for mass conservation.)

Regions C and F (Northern Boundary Region). In these regions, it is assumed that the only physical process of significance is the advection by downwelling water and all particulate fluxes are neglected as being small in comparison. The dissolved phase equation is then

$$\ell \overline{C}'_t + u_o C_o + \ell (w'\overline{C}')_z = -\lambda \ell \overline{C}' + \ell \overline{Q}'_I \qquad \text{(A.IX.14)}$$

which, since if there is no entrainment $u_o = 0$ and $\ell w' = -Lw = $ constant, reduces to

$$\overline{C}'_t - (L/\ell)w\overline{C}'_z = -\lambda \overline{C}' + \overline{Q}'_I \qquad \text{(A.IX.15)}$$

in Region C.

In Region F, it is also assumed that diffusive effects and particulate interactions are unimportant in comparison with advective effects. Thus, since water is leaving this region, $C_o = C'$. Remembering that $u_o = \ell w'_z$, one obtains Equation (A.IX.15) for Region F as well as Region C (although w is not constant).

Regions A and B (Surface Layer). The surface layer, A, which connects the interior to the northern boundary region is taken to be a well-mixed layer of thickness $\hat{\delta}$. All horizontal variations are neglected. The concentrations for the dissolved and particulate phases in this layer are determined by the fluxes in and out of it as well as by the interaction between the dissolved and particulate phases. The most complicated expression is for the dissolved phase and is given by

$$\frac{\partial}{\partial t} (\hat{\delta} \overline{C}_1) = - w\overline{C}_1 \quad + \quad w\overline{C}(0\_) \quad + \quad Q_s \quad - \quad K_V \overline{C}_z(0\_) \quad (\text{A.IX.16})$$

$$\begin{bmatrix} \text{loss to} \\ \text{northern} \\ \text{boundary} \\ \text{region} \end{bmatrix} \quad \begin{bmatrix} \text{upwelling} \\ \text{from} \\ \text{below} \end{bmatrix} \quad \begin{bmatrix} \text{surface} \\ \text{input} \end{bmatrix} \quad \begin{bmatrix} \text{diffusive} \\ \text{exchange} \\ \text{with} \\ \text{Region E} \end{bmatrix}$$

$$- (\lambda + k_1) (\hat{\delta} \overline{C}_1) + \frac{1}{\tau_i} \hat{\delta} \overline{C}_{p_i} + \hat{\delta} \overline{Q}_I$$

$$\begin{bmatrix} \text{loss by} \\ \text{radioactive} \\ \text{decay} \end{bmatrix} \begin{bmatrix} \text{loss to} \\ \text{inorganic} \\ \text{particles} \end{bmatrix} \begin{bmatrix} \text{gain from} \\ \text{dissolving} \\ \text{biological} \\ \text{matter} \end{bmatrix} \begin{bmatrix} \text{gain from} \\ \text{local} \\ \text{interior} \\ \text{source} \end{bmatrix}$$

$$- \mu_i \hat{\delta} \gamma_i \overline{C}_1$$

$$\begin{bmatrix} \text{loss due to} \\ \text{biological uptake} \end{bmatrix}$$

where the subscript 1 indicates the value of the variable in the well-mixed surface layer, and 0_ is used to indicate that the value to be used is that just below the surface layer. μ_i^{-1} is the rate of production of particulate biological matter and τ_i is the time-scale for its dissolution and γ_i is the concentration factor for the biological matter being produced.

The equations for particulate matter are similarly

$$\frac{\partial}{\partial t} (\hat{\delta} \overline{C}_{p_1}) = w_p \overline{C}_{p_1} - K_V \overline{C}_{pz}(0\_) + k_1 (\hat{\delta} C_1) - (\lambda + k_2)(\hat{\delta} \overline{C}_{p_1}) \qquad (\text{A.IX.17})$$

$$\frac{\partial}{\partial t} (\hat{\delta} \overline{C}_{p_{i_1}}) = w_{p_i} \overline{C}_{p_{i_1}} - \lambda \hat{\delta} \overline{C}_{p_{i_1}} - \frac{1}{\tau_i} \hat{\delta} \overline{C}_{p_{i_1}} + \mu_i \hat{\delta} \lambda_i \overline{C}_1 \qquad (\text{A.IX.18})$$

<u>in Region A,</u>

where the diffusive flux of organic matter has been neglected relative to the advective flux in Equation (A.IX.18).

The reader should note that w_{p_i}, μ_i, τ_i, and the proportion of the volume that is occupied by biological matter are related through an equation which describes the rate of change of biological matter, f_i. For the purposes of what follows, however, it is sufficient to choose these parameters in a way which is consistent with the available data.

It is assumed, consistent with the underlying Region C, that particulate processes are unimportant in Region B. Thus the contaminant is simply advected through this region as it decays.

122

The bottom boundary layer. The model as formulated contains a rapidly mixed (presumably by bioturbation) bottom sedimentary layer in which first-order geochemical processes take place and some of the particulate matter reaching the sea floor dissolves releasing the substance it is carrying. If one lets f' be the fraction of the sediment volume occupied by particles, $\overline{C}'_w$ the horizontally averaged concentration per unit volume of water and $\overline{C}'_p$ the corresponding concentration per unit sediment volume, integrating over the sediment layer yields for the pore-water concentration:

$$(1-f')h\overline{C}'_{wt} \quad = \quad hk_2^s f'\overline{C}'_p \quad - \quad h(1-f')\lambda\overline{C}'_w \quad - \quad h(1-f')k_1^s\overline{C}'_w \qquad (A.IX.19)$$

$$\begin{bmatrix} \text{rate of change} \\ \text{of total amount} \\ \text{in water} \end{bmatrix} \quad \begin{bmatrix} \text{gain from} \\ \text{particles} \end{bmatrix} \quad \begin{bmatrix} \text{radioactive} \\ \text{decay} \end{bmatrix} \quad \begin{bmatrix} \text{loss to} \\ \text{particles} \end{bmatrix}$$

$$+ \quad K_v\overline{C}_z(-H_+) \quad - \quad (1-\delta)w_p\overline{C}_p(-H_+) \quad - \quad w_s(1-f')C'_w \quad - \quad w_{p_i}C_{p_i}(-H_+)$$

$$\begin{bmatrix} \text{diffusive} \\ \text{flux across} \\ \text{sediment} \\ \text{surface} \end{bmatrix} \quad \begin{bmatrix} \text{loss from} \\ \text{settling} \\ \text{out of} \\ \text{scavenging} \\ \text{particles} \end{bmatrix} \quad \begin{bmatrix} \text{burial} \end{bmatrix} \quad \begin{bmatrix} \text{loss from} \\ \text{biological} \\ \text{particles} \\ \text{settling} \\ \text{out} \end{bmatrix}$$

where h is the well-mixed sediment layer thickness, k_1^s, and k_2^s are the bulk particle-water first-order rate constants, w_s is the sediment accumulation rate, $(1-\delta)$ is the fraction of scavenging particulate matter that dissolves releasing the substance it carries to the pore water, and H_+ indicates that the value is to be taken just above the sediment layer.

All of the biological matter (C_{p_i}) crossing the sediment interface is also assumed to decompose. We note that it is possible that not all of the contaminant originally on the particles that dissolve is released to the water. If some of the contaminant remains with the undissolved portion, it is possible for equilibrium to be upset in the sediments with more contaminant being held on the particles than would be determined from our model. This would result in a decreased diffusive flux of dissolved contaminant into the water and an increased burial rate on particles. Though our bottom boundary condition could be modified to accommodate this (as well as finite dissolution and diffusion rates in the bioturbated layer and a non-zero diffusive flux to the region below), this possibility has not been pursued.

The corresponding particulate equation is:

$$f'hC'_{pt} = hk_1^s(1-f')\overline{C}'_w - hf'\lambda\overline{C}'_p - hf'k_2^s\overline{C}'_p - \delta w_p\overline{C}_p(-H_+) - w_s f'\overline{C}'_p \qquad (A.IX.20)$$

If f is the fraction of volume occupied by sediment in the water above and if we assume the basic first-order rate constants in the water and

123

sediments are equal, it may be shown that the bulk rate constants in the sediments are given by:

$$k_1^s = (\frac{f'}{1-f'}) \quad (\frac{1-f}{f})k_1$$

$$k_2^s = k_2 \tag{A.IX.21}$$

Also, w_s, the sediment accumulation speed, must be related to the particulate settling speed, w_p, by:

$$- f'w_s = \delta f w_p \tag{A.IX.22}$$

Finally, it is assumed that the pore water in the boundary layer is in equilibrium with the overlying water mass, that is,

$$\overline{C}(-H_+) = \overline{C}_w' \tag{A.IX.23}$$

The equations for the various regions given above now form a closed set, some of the properties of which are discussed below.

The steady-state problem. As discussed earlier, it is instructive to consider the time-independent solution of the problem formulated above and to relate solutions to naturally occurring substances. In this case Equation (A.IX.12) can be solved analytically and the equations in the various regions reduce to:

$$w\overline{C}_z = (K_V\overline{C}_z)_z + k_2\overline{C}_p - (\lambda + k_1)\overline{C} + (1/\tau_i)\overline{C}_{p_i}(0)\exp\left(\left(\frac{1+\lambda\tau_i}{-w_{p_i}\tau_i}\right)z\right) + Q_I \tag{A.IX.24}$$

$$w_p\overline{C}_{pz} = (K_V\overline{C}_{pz})_z + k_1\overline{C} - (\lambda + k_2)\overline{C}_p \tag{A.IX.25}$$

in Region E;

$$- w_z\overline{C}' + (w\overline{C})_z = (K_V\overline{C}_z)_z + k_2\overline{C}_p - (\lambda + k_1)\overline{C} + (1/\tau_i)\overline{C}_{p_i}(0)\exp\left(\left(\frac{1+\lambda\tau_i}{-w_{p_i}\tau_i}\right)z\right) + \overline{Q}_I \tag{A.IX.26}$$

$$w_p\overline{C}_{pz} = (K_V\overline{C}_{pz})_z + k_1\overline{C} - (\lambda + k_2)\overline{C}_p \tag{A.IX.27}$$

in Region D;

$$- (L/\ell)w\overline{C}_z' = -\lambda\overline{C}' + \overline{Q}_I \tag{A.IX.28}$$

in Regions C and F.

For a steady-state balance, the concentration on organic material in the surface layer from Equation (A.IX.18) satisfies

$$(w_{p_i} - \frac{\hat{\delta}}{\tau_i} - \lambda \hat{\delta})\overline{C}_{p_{i_1}} = - \mu_i \hat{\delta} \gamma_i \overline{C}_1 \qquad \text{(A.IX.29)}$$

Since however the fraction of the surface layer volume occupied by organic matter must also be in steady state

$$-(w_{p_i} - \frac{\hat{\delta}}{\tau_{i.}})f_i = \mu_i \hat{\delta} \qquad \text{(A.IX.30)}$$

and thus

$$(\mu_i + \lambda f_i)\overline{C}_{p_{i_1}} = \mu_i f_i \gamma_i \overline{C}_1 \qquad \text{(A.IX.31)}$$

Equation (A.IX.31) shows that there exists an expression of the form $C_{p_i} = \tilde{\gamma} \overline{C}$ but that, even if radioactive decay in the surface layer is negligible, $\tilde{\gamma}_i = f_i \gamma_i$ depends both on the rate of production of biological particles and their concentration factor for the substance being considered. In what follows $\tilde{\gamma}_i$ will be varied to investigate the possible influence of biological particles. The values of $\tilde{\gamma}_i$ used will ultimately have to be consistent with what is known of the biological particle flux $w_{p_i}f_i$ and the concentration factor γ_i.

We shall also in the following analysis assume that scavenging particles are in equilibrium with the water in the surface layer, i.e.

$$k_1\overline{C} = (\lambda + k_2)\overline{C}_p \quad \text{at } z = 0 \qquad \text{(A.IX.32)}$$

Summing Equations (A.IX.16) − (A.IX.18) and recognizing that $\overline{C}_1 = \overline{C}(0_-)$, we have

$$Q_s + \delta Q_I = K_V(\overline{C}_z + \overline{C}_{pz}) - w_p \overline{C}_p - w_{p_i} \overline{C}_{p_i} - \lambda \delta (\overline{C}_1 + \overline{C}_p + \overline{C}_{p_i}) \quad \text{at } z = 0$$

Finally, the approximation has been made that the surface layer is thin, so that

$$Q_s = \delta K_V(\overline{C}_z + C_{pz}) - w_p C_p - w_{p_i} \overline{C}_{p_i} \quad \text{at } z = 0 \qquad \text{(A.IX.33)}$$

125

together with the equations for the surface and bottom boundary layers Equations (A.IX.24) - (A.IX.28) form a closed set. The range of oceanic values for the parameters λ, k_1, k_2, γ_i, μ_i, K_V, etc. yield a number of particular balances between diffusion, advection, particulate flux, radioactive decay, etc.

2. APPLICATION OF THE MODEL TO THE NORTH PACIFIC

In this Section, some features of the model are revealed through its application to the observed profiles of several natural species from the North Pacific. These examples illustrate some of its strengths and some of its weaknesses.

The examples we have chosen to consider are ^{230}Th (interior source, $T_{\frac{1}{2}} = 7.52 \times 10^4$ a, $K_d \sim 2 \times 10^7$), ^{226}Ra (bottom source, $T_{\frac{1}{2}} = 1622$ a, $K_d \sim 10^3$), and ^{210}Pb (surface and interior sources, $T_{\frac{1}{2}} = 21.4$ a, $K_d \sim 8 \times 10^6$) from the radioactive decay series of ^{238}U, and the stable elements Mn($K_d \sim 10^7$), Cu($K_d \sim 1.5 \times 10^6$), Ni($K_d \simeq 0$) and Cd($K_d \simeq 0$) all of which have their source at the surface. These examples cover a wide range of λ and K_d, as well as a variety of source distributions.

In the application of a model such as this to a single ocean basin, one is, as usual, immediately faced with the reality that some of its features are not directly applicable to the basin being considered. Of primary importance in the Pacific is that although upwelling of deep water is believed to occur in the basin, the source waters come from outside and carry with them properties of other basins. Thus, in applying the model one is relying on its "generic" nature to represent properties of the basin that have similarities from basin to basin. Properties that are locally different from the horizontal average are unlikely to be at all well represented.

Choice of parameters. The ocean is taken to be 5000 m deep with a northern boundary region that is one-hundredth of the width of the interior upwelling region. The upwelling velocity is taken to be 1.16×10^{-5} cm s^{-1} or 0.01 m d^{-1}, giving a downwelling velocity in the northern region of 1 m d^{-1}. The solutions are in general insensitive to the choice of width of the northern region since all that is important is that the water (and element) is recirculated back to the bottom region and that its radioactive decay is accounted for.

The bottom boundary region is taken to be 500 m thick. If the concentration levels change little over the depth of the fluid, this choice makes little difference. However, if there is significant variation with depth, then the choice influences the form of the solutions near the bottom. In particular, we note that, although both C and C_z are continuous across the top of the bottom boundary region, water from the northern boundary will force concentration levels in the bottom region towards those at the surface.

The sediment layer is taken to be 10 cm thick and 0.3 of its volume is occupied by particles (i.e. f' = 0.3). Although the shape of the vertical profiles should be insensitive to these choices, if either burial or radioactive decay in the sediment layer is a significant loss term, the absolute concentration levels may be strongly dependent on this choice. If $\lambda = 0$, one can easily see this is so since the only removal mechanism is burial, which is strongly dependent on K_d and f'. In

126

general ($\lambda \neq 0$), the thicker the boundary layer and the greater the fraction which is particulate material, the more of a substance that will be retained there. This effect is of course greatest for large K_d's. More information is required on both f', and the effective thickness, h^*, of the boundary layer that is in contact with the water above (no doubt both are actually highly spatially dependent).

The vertical dispersion coefficient is taken to be 1.16×10^{-4} m^2 s^{-1} (10 m^2 d^{-1}). Though this is generally accepted as a reasonable estimate, it should be noted that K_V may vary substantially through the water column and thus influence the observed vertical profiles when diffusion is important.

The fraction of the water volume occupied by particles, f, is taken to be 10^{-8}, a rough estimate. The more important quantity to get right is the vertical flux of particles, fw_p. This has been estimated (see Figure A.I-2) to be of order 0.5 $mg \cdot cm^{-2}$ a^{-1} in the North Pacific. For the assumed density of particles this gives $w_p = 4.17 \times 10^4$ cm a^{-1} or ~ 1.1 m d^{-1}.

When the vertical flux on particles is large, the solutions will be sensitive to this choice. However, this appears to be the best estimate available and as we shall see, gives results consistent with ^{230}Th profiles where the transport is strongly dominated by the particulate flux.

The parameters which remain to be specified are:

    (1)   the geochemical constants k_1, K_d, λ;
    (2)   the sources;
    (3)   the biological factors $\tilde{\gamma}_i$ w_{p_i} and τ_i; and
    (4) δ, the fraction of scavenging particles that are not decomposed in the sediments.

While (1) and (2) have been estimated using available information, (3) and (4) will be varied to find values consistent with the observations. Table A.IX-I summarizes the various parameter choices considered.

2.1.   Natural radionuclides

^{230}Th (Interior Source). The source of ^{230}Th in the ocean is the radioactive decay of ^{234}U. At and near GEOSECS Station 226, Nozaki et al. (1981) give a uniform source strength of 8.2×10^{-13} dpm kg^{-1} s^{-1} [4] (averaged over a depth of 5000 m this gives 0.354 dpm m^{-2} d^{-1}). Using a scavenging model and the available data for both ^{230}Th and ^{234}Th, they determine a value for k_1 of 1.5 a^{-1} (4.1×10^{-3} d^{-1}) and for k_2, 6.3 a^{-1} (1.7×10^{-2} d^{-1}). In the following, k_1 is taken to have this value for all elements, as is consistent with the idea that k_1 is independent of the element under consideration and determined by the availability of surface sites on particulate matter. k_2 is determined from $k_2 = k_1/fK_d$. For ^{230}Th this gives $k_2 = 7.5$ a^{-1}. The difference from the value of 6.3 a^{-1} determined by Nozaki et al. could be compensated for by a different choice of f. This has not been done

---

[4] dpm = disintegrations per minute.

TABLE A.IX-I. PARAMETERS FOR VARIOUS RADIOACTIVE AND STABLE NUCLIDES

Fixed parameters:

$H = 5000$ m $W_{max} = 0.01$ m·d^{-1} $K_1 = 4.1 \times 10^{-3}$·d^{-1}
$Z_1 = 500$ m $W_p = -1.1$ m·d^{-1} $f = 10^{-8}$
$\ell/L = 0.01$ $W_{p_i} = -10$ m·d^{-1} $f' = 0.3$
$h = 0.1$ m $w_s = -(\delta f/f')w_p = 2.75 \times 10^{-8}$ m·d^{-1} (for $\delta = 0.75$)

Varying parameters:

Figure	A.IX-2	A.IX-3	A.IX-4	A.IX-5	A.IX-6	A.IX-7
Element	Thorium-230	Radium-226	Lead-210	Manganese	Copper	Nickel and Cadmium
Half-life	7.52×10^4 a	1622 a	21.4 a	Stable	Stable	Stable
K_d (water column)	2×10^7	10^3	8×10^6	10^7	1.5×10^6	0
K_d^s (sediment layer)	2×10^7	10^3	8×10^6 in general 2×10^5 for dotted curve	10^7	1.5×10^6	2×10^5
δ	0.50, 0.75, 1.00	0.75	0.75	0.50, 0.75	0.75	0.75
Sources	Uniform interior source	Bottom source and surface source of 1/10 strength	Interior source as illustrated and a surface source of 0.3 units m^{-2}·d^{-1}	Surface source	Surface source	Surface source
K_V (Region E)	10 m^2·d^{-1}	10 m^2·d^{-1}	10 m^2·d^{-1}	10 m^2·d^{-1}	10 m^2·d^{-1}	10 m^2·d^{-1}
K_V (Region D)	100 m^2·d^{-1} for dotted curve 10 m^2·d^{-1} for others	100 m^2·d^{-1} for dotted curve 10 m^2·d^{-1} for others	100 m^2·d^{-1} for dotted curve 10 m^2·d^{-1} for others	100 m^2·d^{-1} for dotted curve 10 m^2·d^{-1} for others	100 m^2·d^{-1} for dotted curve 10 m^2·d^{-1} for others	100 m^2·d^{-1} for all broken curves 10 m^2·d^{-1} for others
Biological parameters						
γ_1, τ_1	(0., −)	(0.,−) solid (0.003, 500 d) others	(0., −) ✗✗✗ (0.003, 500 d) others	(0., −) solid curves (0.01, 5000 d) others	(0., −) solid curve (0.003, 500 d) others	(0., −) ✗✗✗ (0.005, 2000 d) others
γ_2, τ_2	(0., −)	(0.,−)	(0., −)	(0., −)	(0., −)	(0., −) ✗✗✗ ○-○-○ (0.005, 50 d) ----- (0.01, 50 d) ....

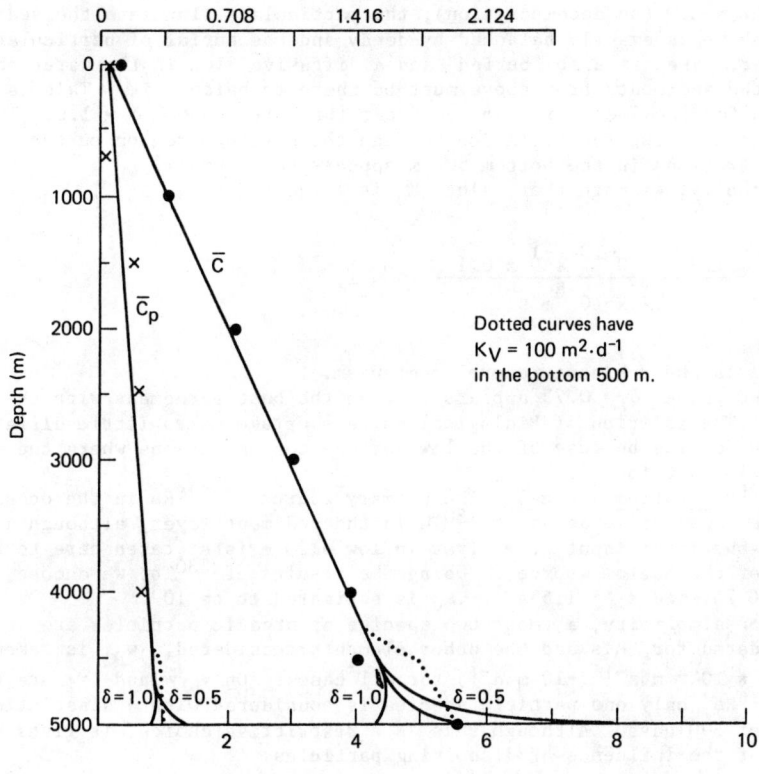

Concentration (dis/min·m^{-3}) for a source
strength of 0.354 dis/min·m^{-2}·d^{-1}

Dotted curves have
$K_V = 100$ m^2·d^{-1}
in the bottom 500 m.

Concentration (units·m^{-3}) – unit source m^{-2}·d^{-1}

^{230}Th: $\lambda = 2.52 \times 10^{-8}$·d^{-1}, $K_d = 2.0 \times 10^7$, interior source.

FIG. A.IX-2. *Model and data comparisons for* ^{230}Th *(see text for details).*

since: (1) the difference appears to be within the uncertainties, and
(2) even though both f and k_2 may be in error by a small amount, the
two important qualities, particulate flux and K_d, have been reasonably
estimated so that the particulate flux of the element being considered
will also be reasonably estimated.

The quantities γ_i, w_{p_i}, τ_i and δ remain to be specified.
Initially we take $\gamma_i = 0$, i.e. biological particles do not carry the
element, and vary δ between 0 and 1. The results of this are
illustrated in Figure A.IX-2. The agreement with Figure A.I-5 is very
good.

The work of Nozaki et al. (1981) indicates that the fluid in the
interior is found to be very near equilibrium ($k_1\bar{C} = k_2\bar{C}_p$) with
local production being balanced by the divergence of the particulate flux

$(Q_1 \simeq w_p C_{pz})$. Also as expected, as the decomposition of
particulates in the sediment layer increases, the bottom concentrations
increase as a result of decreased particulate flux out of the sediments.
When $\delta = 1.0$ (no decomposition), the particulate flux into the sediments
from above is exactly balanced by decay and the burial of particulates.
However, water is also "buried" and a diffusive flux in the water phase
into the sediments from above must be there to balance it. This is the
reason for the local minimum seen near the bottom when $\delta = 1.0$. The
influence of the recirculation through the northern region on the
concentrations in the bottom 500 m appears to be small.

   Finally, we note that, since K_d is large and

$$\frac{\lambda h}{w_s} \sim \frac{2.93 \times 10^{-13} \text{ s}^{-1} \times 0.1 \text{ m}}{3.7 \times 10^{-8} \text{ m s}^{-1}} \ll 1,$$

burial is the dominant removal mechanism.
   The value $\delta = 0.75$ appears to give the best agreement with the
data. The addition of biological surface uptake makes little difference
to the results because of the low surface concentrations where the
particles are formed.
   ^{226}Ra (Bottom Source). The primary source of ^{226}Ra in the ocean
is the radioactive decay of ^{230}Th in the sediment layer, although a
small surficial input from river inflow also exists (taken here to be
1/10 of the bottom source). Using the results of ^{230}Th, we choose
$\delta = 0.75$, and $k_1 = 1.5 \text{ a}^{-1}$. K_d is estimated to be 10^3.
   For simplicity, at most two species of organic particles are
considered for this and the other elements considered. w_{p_i} is taken as
$-1.16 \times 10^{-4} \text{ m s}^{-1}$ (-10 m d^{-1}) for all cases. Only $\tilde{\gamma}$ and τ_i are varied.
For ^{226}Ra only one particle species is considered with a dissolution
time of 500 days. Although this is a restrictive choice, it gives some
idea of the influence of dissolving particles.

   The results are shown in Figure A.IX-3. Due to the small K_d ($\sim 10^3$)
for ^{226}Ra, the scavenging particles play little role. When biological
particles are not included, the element is transported into the interior
of the fluid by diffusion from both the surface and the bottom and in the
interior, advection and diffusion are balanced primarily by radioactive
decay.
   When biological particles are included, ^{226}Ra is rapidly
transported from the surface into the interior. Some dissolves in the
interior and some reaches the bottom, where it dissolves and then
diffuses back into the water column. Although the shape of the
concentration profiles is strongly affected by the presence of dissolving
particles, the mean concentration in the water column changes little as
radioactive decay remains the primary sink, and K_d is sufficiently
small that the sediments play only a secondary role. That is,

$$\left[\frac{(\lambda h - w_s) f' K_d}{\lambda H} \sim \frac{h f' K_d}{H} \sim 0.006 \right]$$

^{210}Pb (Surface and Interior Sources). The major source of ^{210}Pb
in the ocean is the radioactive decay of ^{226}Ra although there is
significant surface input. Due to the small K_d of ^{226}Ra the relative

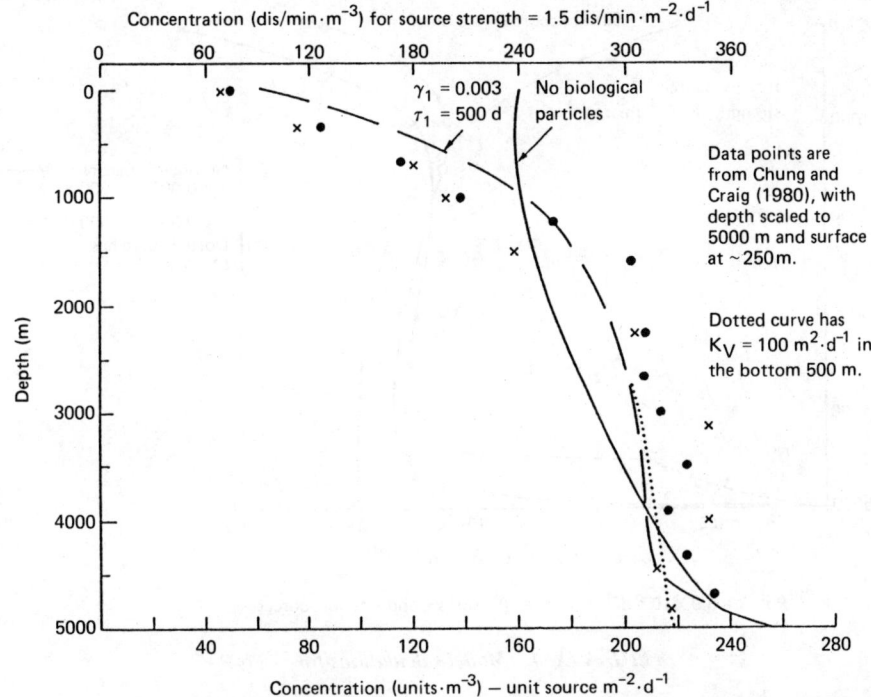

Concentration (dis/min·m^{-3}) for source strength = 1.5 dis/min·m^{-2}·d^{-1}

γ_1 = 0.003
τ_1 = 500 d

No biological particles

Data points are from Chung and Craig (1980), with depth scaled to 5000 m and surface at ~250 m.

Dotted curve has K_V = 100 m^2·d^{-1} in the bottom 500 m.

Depth (m)

Concentration (units·m^{-3}) — unit source m^{-2}·d^{-1}

^{226}Ra: λ = 1.17 $\times$ 10^{-6}·d^{-1}, K_d = 10^3, bottom source and surface source of 1/10 strength

Data points from Chung and Craig (1980).

FIG. A.IX-3. Model and data comparisons for ^{226}Ra.

contribution from the sediments is less than 0.01 of the contribution from above. Thus, using the above calculated distribution of ^{226}Ra one obtains a depth-dependent internal source for ^{226}Ra as illustrated in Figure A.IX-4 and a bottom source of 0.006 units m^{-2}. The surface input is taken to be 0.3 units m^{-2}. At most one type of biological particle is used and it is taken to have the same properties as for ^{226}Ra.

The model results are shown in Figure A.IX-4. The solid curve marked by crosses corresponds to K_d = 8 x 10^6, no biological uptake in the surface layer, h = 3 cm, and all other chemical and physical parameters as determined previously.

The value of 3 cm for h used here recognizes that given the short half-life of ^{210}Pb it will not be well-mixed through the sediments. The effective thickness is then given by $(K_b/\lambda)^{\frac{1}{2}}$ (see Appendix V) where K_b has been taken as 10^{-12} m^2 s^{-1}.

The results show a strong surface maximum and a rapid decrease in concentration at the bottom. The surface maximum is reduced if surface uptake into organic particles is included. However, the bottom concentration remains small even if K_V is increased by a factor of 10

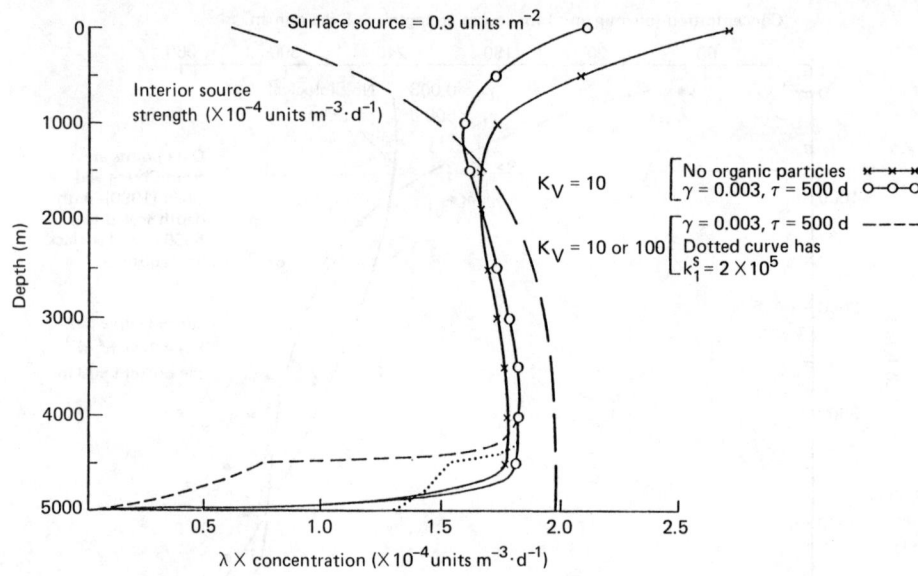

FIG. A.IX-4. *Model calculations for* ^{210}Pb.

(to 100 m^2 d^{-1}) in the bottom 500 m (the broken curve). Indeed if
one lets Q_{sed} be the total source of ^{210}Pb to the upper 3 cm of
sediment, then for steady state $Q_{sed} \simeq \lambda hf'K_d$ C(-H). Taking
$\overline{C}$ (-H) = 200 dpm m^{-3}, as estimated by Bacon et al. (1976), this gives
$Q_{sed} \sim 1.3$ x 10^3 dpm m^{-2} d^{-1}. Comparing this with the total rate of
supply from ^{226}Ra of about 1.6 x 10^2 dpm m^{-2} d^{-1} from both the surface
and the column sources, it is clear that even the addition of a
significant bottom source could not maintain the estimated sedimentary
inventory of ^{210}Pb. Assuming that C(-h) is not grossly in error, it is
apparent that the quantity $hf'K_d$ has been overestimated. An example
of the influence of reducing this quantity (for example, by reducing K_d
in the sediment layer) by a factor of 0.025 is given by the dotted curve
in Figure A.IX-4.

Finally, we note that regardless of assumptions made about the
sediment layer the balance in region E below 1500 m is primarily between
the local source, radioactive decay and removal on inorganic particles.
Thus, the local loss due to particulate settling reduces the
concentration below what would be observed if the local source were
balanced by radioactive decay alone. This is clearly in agreement with
the observations (see Figure A.I-3b).

2.2.   Stable elements

Mn (Surface Source). Surficial input is the main source of manganese
in the ocean. Profiles of this element have a distinct maximum at the
bottom, as illustrated in Figure A.I-3b. This feature is caused by the

132

interior particulate scavenging and transport to the bottom resulting
from the large K_d ($\sim 10^7$) of this element. Model results are shown
in Figure A.IX-5. Without biological uptake a strong bottom maximum with
relatively uniform upper ocean values can be produced as long as there is
release of Mn from particulate matter at the bottom (δ small). The same
result can also be obtained with surface biological uptake onto slowly
dissolving particles ($\tau_1 \sim 5000$ days) and little bottom release from
scavenging particles at the bottom ($\delta \simeq 1$) is included. The near-surface
maximum is thought to result from a change in the redox chemistry of
magnesium at the depth of the O_2 minimum.

Cu (Surface Source). As for manganese, the major source of copper
in the ocean is through the surface, and again the maximum observed
concentrations exist at the bottom (Figure A.I-3b). Copper profiles
however show a continuous variation in concentration from the surface to
the bottom. If surface biological uptake is not included, the model
gives profiles similar to those for manganese with relatively uniform
interior concentrations enhanced at the bottom due to the relatively
large K_d ($\sim 1.5 \times 10^6$). However, the inclusion of a small amount of
biological uptake in the surface layer onto particles with release into
the interior at a decay scale of the same order as the depth gives
profiles similar to those observed (Figure A.IX-6).

Ni and Cd (Surface Source). Nickel and cadmium also have their major
inputs at the ocean surface and are relatively non-reactive in the water
column ($K_d \simeq 0$). If these elements were similarly non-reactive in the
sediments, their residence time in the water column would be of order
5×10^8 a. However, Sclater et al. (1975) have estimated the residence
time for Ni to be of order 10 000 a, which can only be the case if it is
associated with particles in the sediment layer. To account for
this within the context of the present model, an appropriate value of
K_d for the sediment layer (K_d^s) must be chosen. This may be done
by taking $H/(w_s f' K_d^s) \sim 10\ 000$ a or $K_d^s \sim 1.66 \times 10^5$. The results shown in
Figure A.IX-7 are for $K_d^s = 2 \times 10^5$.

The solid line marked by crosses shows the nearly uniform
distribution that would be obtained in the absence of biological uptake
in the surface layer. However, biological uptake can alter these
profiles to be similar to those observed. The data also show that,
although the profiles of Cd and Ni are similar, there are qualitative
differences between them. In particular, Cd has a subsurface maximum at
a depth of order 1000 m which is not present in the Ni profile.
Comparison with the profiles of nitrate and silicate suggests that
cadmium is more efficiently taken up by the soft parts (with which
nitrate is associated) of organic matter than is nickel and more rapidly
released into the water column, thus producing the subsurface
concentration maximum. A couple of examples with different amounts of
uptake onto rapidly dissolving particles are shown in Figure A.IX-7 to
illustrate this effect.

The need to rationalize the Cu and Ni profiles in this way
effectively emphasizes how important the details of the interactions,
uptake, and release of various elements are to their distribution. Thus,
simple models with one or two types of particles must often be unable to
represent the transport of a particular substance unless the particulate
interactions are "tuned" to that substance. The need for real data and
understanding of the geochemistry involved is readily apparent.

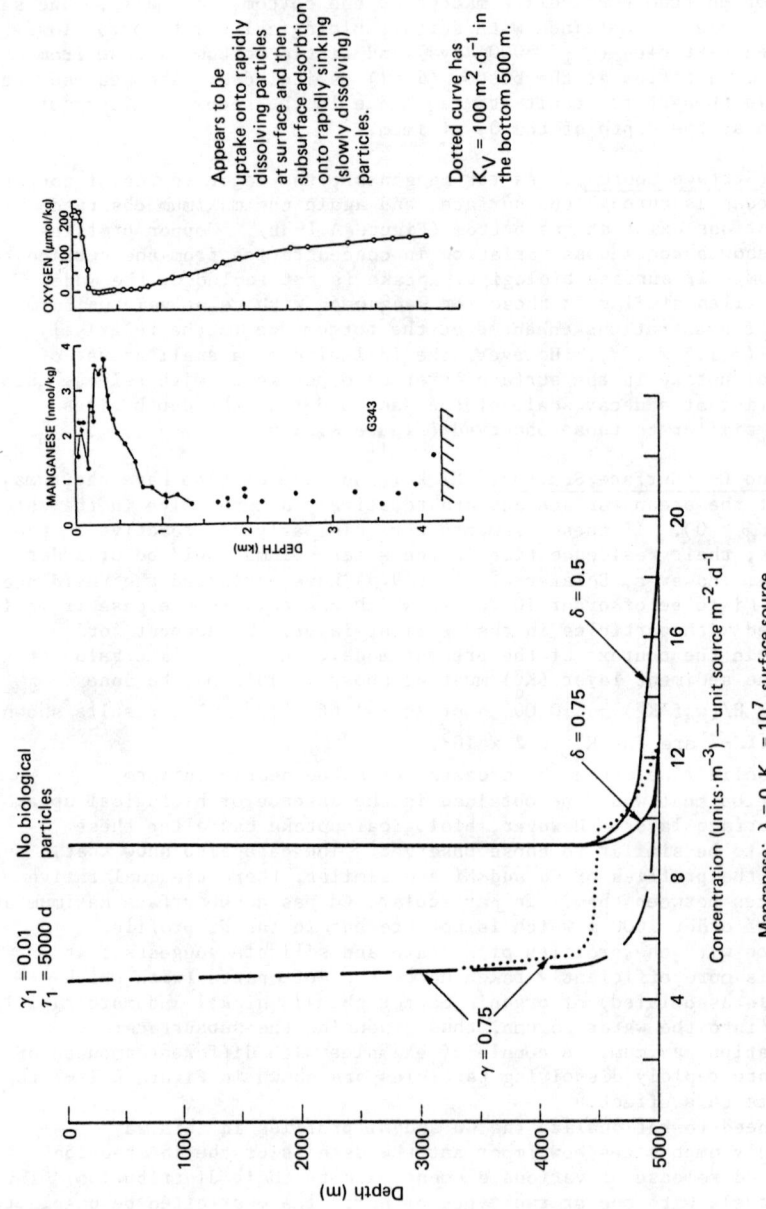

FIG. A.IX-5. Model calculations for manganese.

Appears to be uptake onto rapidly dissolving particles at surface and then subsurface adsorbtion onto rapidly sinking (slowly dissolving) particles.

Dotted curve has $K_V = 100$ m^2·d^{-1} in the bottom 500 m.

No biological particles

$\gamma_1 = 0.01$
$\tau_1 = 5000$ d

$\gamma = 0.75$

$\delta = 0.75$

$\delta = 0.5$

Concentration (units·m^{-3}) — unit source m^{-2}·d^{-1}

Manganese: $\lambda = 0$, $K_d = 10^7$, surface source

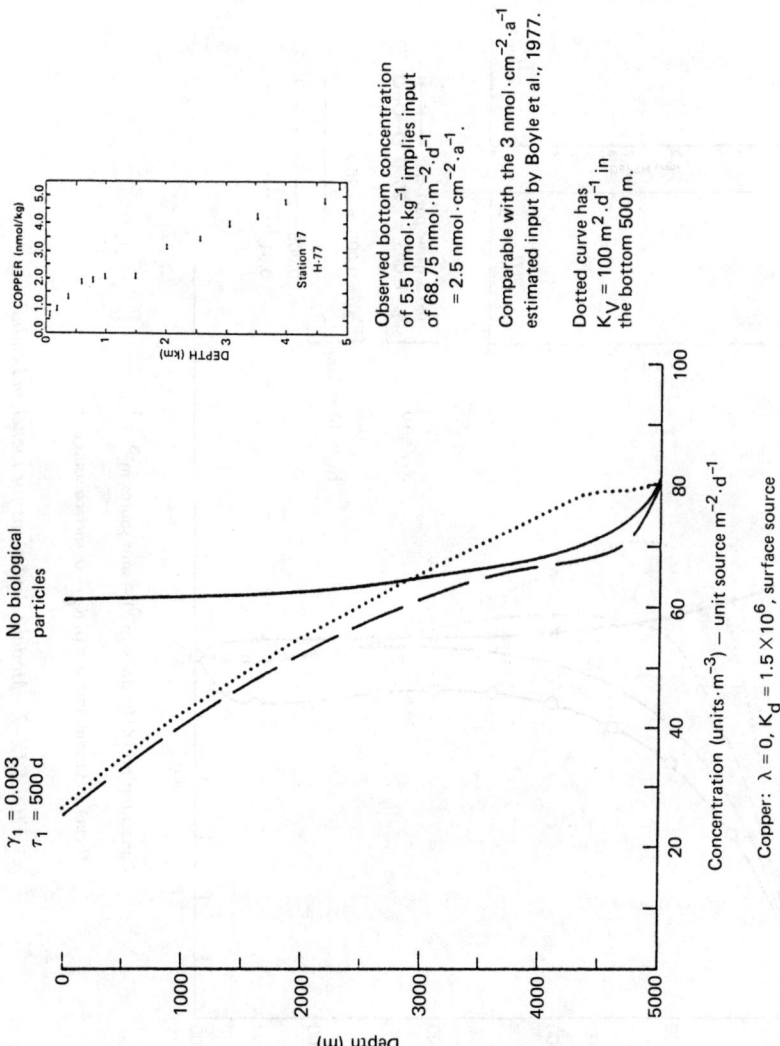

FIG. A.IX-6. Model calculations for copper.

No biological particles

$\gamma_1 = 0.003$
$\tau_1 = 500$ d

Observed bottom concentration of 5.5 nmol·kg^{-1} implies input of 68.75 nmol·m^{-2}·d^{-1} = 2.5 nmol·cm^{-2}·a^{-1}.

Comparable with the 3 nmol·cm^{-2}·a^{-1} estimated input by Boyle et al., 1977.

Dotted curve has $K_V = 100$ m^2·d^{-1} in the bottom 500 m.

Concentration (units·m^{-3}) — unit source m^{-2}·d^{-1}

Copper: $\lambda = 0$, $K_d = 1.5 \times 10^6$, surface source

Depth (m)

COPPER (nmol/kg)

DEPTH (km)

Station 17
H-77

135

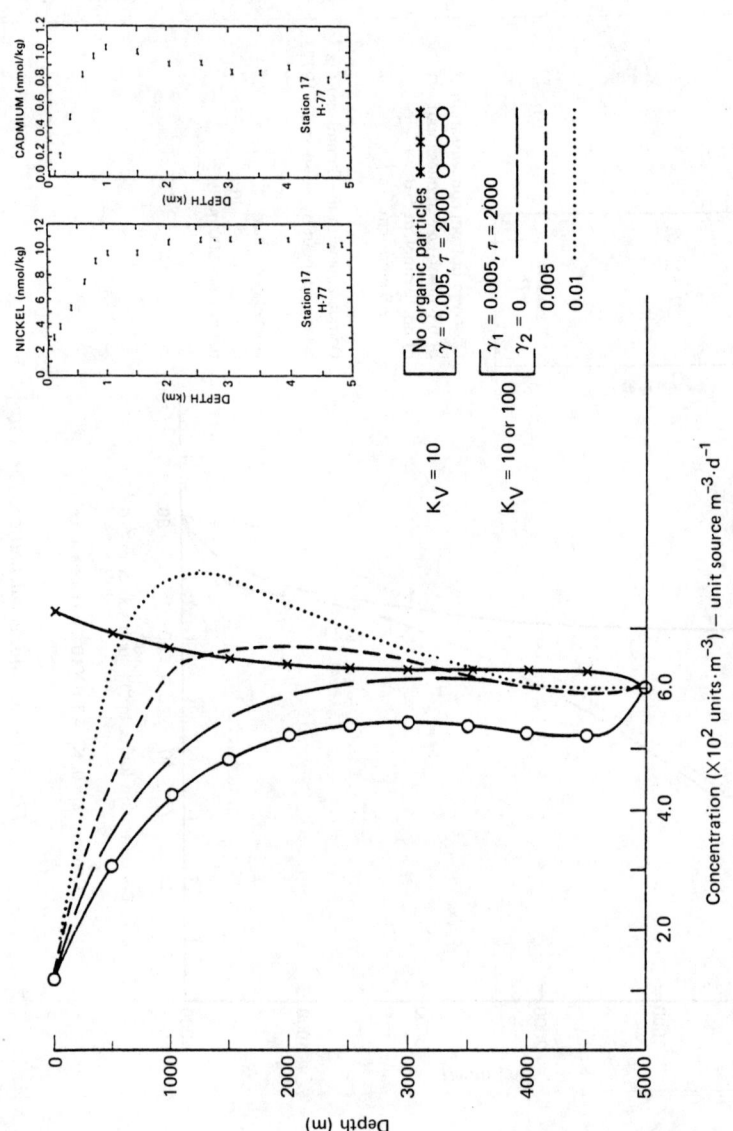

FIG. A.IX-7. *Model calculations for nickel and cadmium.*

Concentration (×10² units·m⁻³) — unit source m⁻³·d⁻¹

Nickel and cadmium: $\lambda = 0$, $K_d \cong 0$, surface source

$K_V = 10$

No organic particles ✱–✱–✱
$\gamma = 0.005$, $\tau = 2000$ O–O–O

$K_V = 10$ or 100

$\gamma_1 = 0.005$, $\tau = 2000$
$\gamma_2 = 0$ ————

0.005 — — —

0.01 ·········

Notwithstanding the problems with Cu and Ni, it is worth noting that most
of the features of the elements considered have been reproduced with a
limited number of particle types with fixed properties.

3. APPLICATION TO CONTAMINANTS WITH BOTTOM SOURCES

The results described above show the ability of the model as
formulated to mimic known geochemical processes as well as the
distributions of naturally occurring substances. However, especially
when biological particles are involved, it has been shown that it is only
capable of reproducing the details of observations when various
parameters are carefully selected for the particular case at hand. This
is not surprising, considering the complexity of the processes that are
being modelled, but shows that the utility of the model as a predictor of
the water concentration of a particular contaminant is dependent on the
role played by particles and one's knowledge of their proper
parametrization.
The model as presented includes sedimentary and water column
processes not found in models in the literature or presented in other
appendices of this report. In keeping with the general approach of this
report that the simplest models consistent with the information available
should be used, it will often not be appropriate to use the full model
for a particular contaminant release scenario. The model does however
provide a framework for examining the importance of certain processes and
for determining which must be eventually included. In the following,
some general results are presented for the case of a bottom source. One
aspect that is examined in some detail is the validity of the assumption
made elsewhere that contaminant concentration of the particulate phase is
in equilibrium with the water phase.

Condition for equilibrium. Let us first investigate the
possibility of simplifying our system of equations by taking

$$\overline{C}_p = \alpha \, \overline{C} \tag{A.IX.34}$$

where α_- is a constant. One can note that from Equations (A.IX.25) and
(A.IX.27), if $\overline{C}$ and $\overline{C}_p$ vary sufficiently slowly with depth,

$$\overline{C}_p \simeq [k_1/(\lambda + k_2)]\overline{C} \tag{A.IX.35}$$

More specifically, if H* is the length scale of variations in C and C_p,
Equation (A.IX.35) is satisfied if

$$\frac{w_p}{(\lambda + k_2)H^*} \quad \text{and} \quad \frac{K_V}{(\lambda + k_2)H^{*2}} \ll 1 \tag{A.IX.36}$$

Taking $w_p = -1$ m d^{-1} and $K_V = 10$ m^2 d^{-1}, the ratio of the
first expression in Equation (A.IX.36) to the second expression,
$(|w_p| H^*)/K_V$, is much greater than one unless H* < 10 m. In this

case however (unless the substance decayed on a time-scale of 10 days invalidating equilibrium directly) there would be a balance near the bottom between diffusion in the water phase and advection on the particulate phase, so that

$$H^* \sim \frac{K_V(\lambda + k_2)}{|w_p| \, k_1}$$

(A.IX.37)

Noting that for reasonable particle densities $\overline{C}_p \ll \overline{C}$ for all but the most reactive elements, one obtains from Equation (A.IX.35) that $k_1/(\lambda + k_2) \ll 1$ and thus that the ratio $|w_p| \, H^*/K_V$ is greater than one even for small H^*. Thus, the condition for equilibrium Equation (A.IX.36) reduces to

$$\frac{|w_p|}{(\lambda + k_2)} \ll H^*$$

(A.IX.38)

This condition can be rewritten as

$$(w_p \overline{C}_{pz})/\overline{C} \ll k_1$$

(A.IX.39)

and may be interpreted as stating that the particles must adjust to equilibrium in a time that is short compared to the time taken for particles to fall through a significant change in $\overline{C}_p$.

Vertical scales. Assuming for the moment that equilibrium is satisfied, the vertical scales $\overline{C}_p$ and $\overline{C}$ will be determined. Later the results are checked against Equation (A.IX.38) to see if this assumption is valid and a specific condition for equilibrium is given, Equation (A.IX.54), for a sustained bottom release.

Neglecting biological uptake at the surface ($\overline{C}_{p_i}(0) = 0$) and interior sources, the addition of Equation (A.IX.11) to Equation (A.IX.10) or (A.IX.13) yields (assuming equilibrium and a steady state)

$$[K_V(1+\alpha)\overline{C}_z - (w + \alpha w_p)\overline{C}]_z - \lambda(1+\alpha)\overline{C} + w_z \overline{C}' = 0$$

(A.IX.40)

where

$$\alpha = \frac{\overline{C}_p}{\overline{C}} = \frac{k_1}{(\lambda + k_2)}$$

(A.IX.41)

in both regions D and E.

Since any substance that reaches the surface in significant quantities will undergo negligible decay during its fast passage through the northern boundary layer, $\overline{C}' \simeq C(0)$ and Equation (A.IX.40) may be approximated by

$$K_V^* \overline{C}_{zz} - w^* \overline{C}_z - \lambda^* \overline{C} = -w_z \overline{C}(0)$$

(A.IX.42)

where $K_V^* = K_V(1+\alpha)$, $w^* = (w+\alpha w_p)$, and $\lambda^* = \lambda(1+\alpha)+w_z$ (A.IX.43)

The solutions of Equation (A.IX.42) in either region D or E (where w_z is constant) are given by

$$\overline{C} = (w_z/\lambda^*)\overline{C}(0)+\widetilde{C} \qquad\qquad\qquad (A.IX.44)$$

where $\widetilde{C}$ satisfies

$$K_V^*\widetilde{C}_{zz} - w^*\widetilde{C}_z - \lambda^*\widetilde{C} = 0 \qquad\qquad\qquad (A.IX.45)$$

Substituting

$$\widetilde{C} = A \exp(\beta z) \qquad\qquad\qquad (A.IX.46)$$

into Equation (A.IX.45) yields

$$K_V^*\beta^2 - w^*\beta - \lambda^* = 0 \qquad\qquad\qquad (A.IX.47)$$

for the local vertical scale, β^{-1}, of $\widetilde{C}$.

For w constant (region E), Equation (A.IX.46) is an exact solution of Equation (A.IX.42) whereas, for w not constant (region D), Equation (A.IX.47) merely gives an estimate of the scale of variations in $\widetilde{C}$.

Noting that for a bottom release $\overline{C}(-H) > \overline{C}(0) \geqslant (w_z/\lambda^*)C(0)$ so that $\widetilde{C} > 0$, one sees that the scale of variation in $\overline{C}$, given by

$$H^* = \overline{C}/\overline{C}_z = ((\frac{w_z}{\lambda^*})\overline{C}(0) + C) \geqslant \beta^{-1} \qquad\qquad\qquad (A.IX.48)$$

From Equation (A.IX.47), one obtains

$$= [w^* \pm (w^{*2}+4K_V^*\lambda^*)^{\frac{1}{2}}]/2K_V^* \qquad\qquad\qquad (A.IX.49)$$

In general, both of these roots will be required to construct a solution to the problem. However, for a bottom release with rapid decay away from the bottom, β must have a negative sign and the following limits are immediately apparent.

$$\beta^{-1} = -(K_V^*/\lambda^*)^{\frac{1}{2}} \quad \text{if} \quad |w^*| << 2(K_V^*\lambda^*)^{\frac{1}{2}} \qquad\qquad (A.IX.50)$$

(decay balancing diffusion)

$$\beta^{-1} = K_V^*/w^* \quad \text{if} \quad w^* << -2(K_V^*\lambda^*)^{\frac{1}{2}} \qquad\qquad (A.IX.51)$$

(particulate transport balancing diffusion)

$$\beta^{-1} = -w^*/\lambda \quad \text{if} \quad w^* >> 2(K_V^*\lambda^*)^{\frac{1}{2}} \qquad\qquad (A.IX.52)$$

(upwelling balancing decay)

Equilibrium (continued). At this point it is useful to combine Equations (A.IX.38) and (A.IX.49) to obtain a sufficient condition for equilibrium for the case of a bottom release. The condition is simply

$$\frac{w_p \, |\beta|}{(\lambda + k_2)} \ll 1 \tag{A.IX.53}$$

where

$$\beta = \{(w + \alpha w_p) - [(w + \alpha w_p)^2 + 4K_V(\lambda(1+\alpha) + w_z)(1+\alpha)]^{\frac{1}{2}}\}/2K_V(1+\alpha)$$

$$\simeq \{(w + \alpha w_p) - [(w + \alpha w_p)^2 + 4K_V(\lambda + w_z)]^{\frac{1}{2}}\}/2K_V \tag{A.IX.54}$$

If Equation (A.IX.53) is not satisfied throughout the water column, equilibrium is not everywhere a valid approximation. When this is the case the concentration levels calculated assuming equilibrium may be seriously in error and the situation will warrant careful consideration.

Concentration levels. Given an estimate of the decay scale, β^{-1}, a simple formula for the approximate bottom concentration can be determined. Noting that input must be balanced by burial and radioactive decay, one obtains

$$\overline{Q} = (\lambda h + w_s)[f'\overline{C}'_p + (1-f')C'_w] + \lambda \int_{-H}^{0} (\overline{C} + \overline{C}_p)dz \tag{A.IX.55}$$

If $\lambda \ll k_2$ (so that equilibrium gives $\overline{C}'_p = K_d\overline{C}'_w$ and $\overline{C}_p = fK_d\overline{C}$) and if β can be reasonably approximated by a constant over the depth for which $\overline{C}$ has a significant value, then

$$\overline{Q} \simeq (\lambda h + w_s)[f'K_d + (1-f')]\overline{C}(-H) + \lambda \left[\int_{-H}^{0} \exp[-\beta(z+H)]dz\right](1+fK_d)\overline{C}(-H)$$

$$\simeq \left| (\lambda h + w_s)[f'K_d + (1-f')] + \lambda[1 - \exp(-\beta H)]/\beta \right| \overline{C}(-H) \tag{A.IX.56}$$

In the limit, $\beta H \ll 1$

$$\overline{C}(-H) \simeq \left| (\lambda h + w_s)[f'K_d + 1 - f'] + \lambda H \right|^{-1} \overline{Q} \tag{A.IX.57}$$

If $\beta H \ll 1$ (actually >2)

$$\overline{C}(-H) \simeq \left| (\lambda h + w_s)[f'K_d + 1 - f'] + \lambda/\beta \right|^{-1} \overline{Q} \tag{A.IX.58}$$

When $\lambda = 0$, Equations (A.IX.57) and (A.IX.58) are identical and as accurate as the assumption of equilibrium in the sediment layer allows.

Sample model results: As an illustration of the ideas discussed above, the full model equations have been integrated using the same parameters as defined earlier in its application to North Pacific geochemical data. A unit bottom source (1 unit $m^{-2} d^{-1}$) has been used with varying λ and K_d. The results are illustrated in Figure A.IX-8. The contour intervals are given in units m^{-3}.

First consider the limit $\lambda = 0$. In this case the vertical nature of the concentration field depends strongly on whether the net transport (w^*) is dominated by upwelling of the water phase and thus positive, or by the downward flux of the particulate phase and thus negative. Given the parameters used, w^* is negative or positive above $z = -H+z_1$ depending on whether K_d is greater or less than 10^6. Thus, for $K_d < 10^6$, $w^* > 0$ in the interior, and since diffusion and advection both carry the contaminant upward, a relatively uniform interior concentration results. For $K_d = 10^6$, $w^* = 0$ in the interior and the only flux there is by diffusion. Since this flux must be constant, C_z is constant for $z > -H+z_1$ and the concentration decreases linearly toward the surface but significant levels of the contaminant are maintained throughout the water column. For $K_d > 10^6$, $w^* < 0$ and since the particulate flux dominates upwelling everywhere the concentration field decreases away from the bottom. One can note however that for $K_d > 10^6$ the equilibrium condition Equation (A.IX.63) may not be satisfied (see Figure A.IX-8f) and thus Equation (A.IX.49) gives only a crude estimate of β. However, since the only sink is the sediments, Equation (A.IX.58) still gives a reasonable estimate of $\overline{C}(-H)$. Indeed, for $\lambda = 0$, burial in the sediments is the only removal mechanism in the model. Equations (A.IX.55), (A.IX.57), and (A.IX.58) all reduce to the same simple exact expression for all K_d's, namely

$$\overline{C}(-H) = \left| w_s (1 - f' + f'K_d) \right|^{-1} Q$$

The concentration values for $\lambda = 0$ are not shown in Figure A.IX-8 but for large K_d's (greater than 10^3) they are essentially equal to those for the smallest λ values given. For smaller K_d's the concentration field is essentially uniform and equal to the bottom value, which may be obtained from the formula given above. It is worth noting that, even for small K_d's, burial in the sediments limits water concentrations in this model.

For $\lambda = 0$, Equations (A.IX.49) – (A.IX.52) can be used to estimate the vertical change in C and the above discussion can be extended to include the effects of decay. The major changes are simply a decrease in the total inventory and an increase in the decline in $\overline{C}$ away from the bottom.

Figure A.IX-8 also shows the accuracy of the equilibrium assumption. Only the bottom is shown since this is where the concentration gradients are greatest and disequilibrium is most likely to occur. One can see that, as is to be expected, equilibrium holds except for the most reactive cases (K_d large). One should note however that as is shown by Equation (A.IX.39), any reduction in k_1, while keeping $k_1/k_2 = fK_d$ constant, will increase the range of λ and K_d over which equilibrium does not hold. As mentioned earlier, disequilibrium does not necessarily cause large errors in the water phase concentration field or the inventory of a contaminant obtained from a model assuming equilibrium. Indeed, since at least for long-lived contaminants the sedimentary concentration must be that necessary to balance the input by burial, the bottom water phase concentration must be almost independent of the degree of equilibrium. Changes in the inventory thus will arise

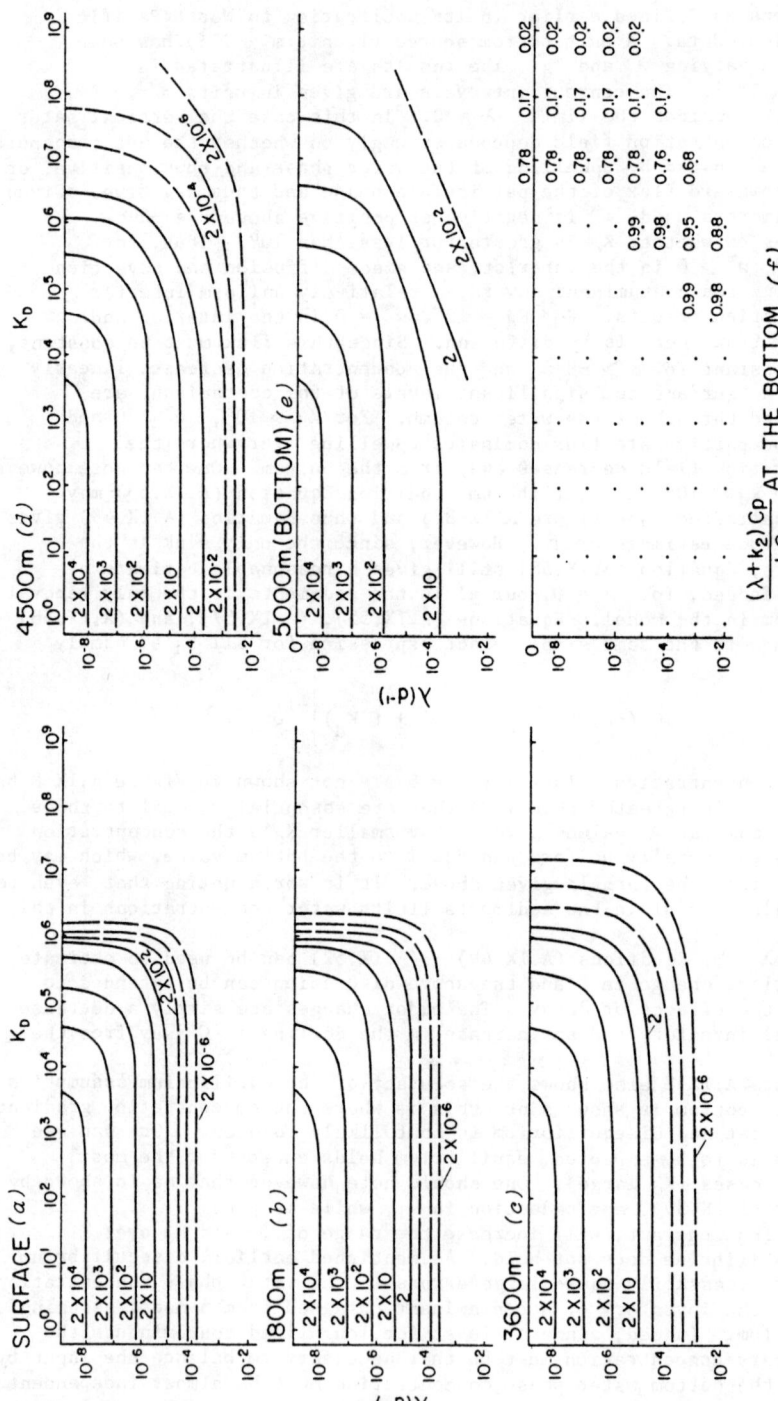

FIG. A.IX–8. Contours of constant concentration as a function of λ and K_d.

only from changes in the vertical scale in the water column and from less than equilibrium concentrations on the sinking particles. In general, the effects of disequilibrium will require careful investigation and, especially for short-lived contaminants, may be dependent on the details of the sedimentary transfer mechanisms and geochemical reactions.

4. CONCLUSION

The model given above provides a self-consistent formulation of some of the physical and geochemical processes that must be considered when estimating the vertical concentration field of a contaminant and has the advantage of being able to reproduce some aspects of the distribution of naturally occurring substances. It can also be easily extended or modified to include more complicated and/or different transport mechanisms. This could be particularly important regarding the sedimentary layer where the model shows the importance for reactive substances of accurately representing geochemical processes.

It is also clear however that many of the processes included are poorly understood and that some of the parameters that are needed have not been experimentally determined. Thus, a model of this complexity is in many ways a research tool with which to test new ideas or hypotheses or to check the importance of processes not included in simpler models. This model or others of a similar nature that could easily be formulated should only be used keeping the above considerations and limitations in mind.

Appendix X

ESTIMATION OF CONCENTRATIONS IN FOOD CHAINS

The calculation of the concentration field of a contaminant in the marine environment need include specific consideration of transport by biological processes only if these are significant compared with parallel physical and chemical processes. The hazards to man and other organisms may nevertheless arise because of exposure via pathways (e.g. food chains) which, whilst trivial in mass transport terms, create a specific linkage between a region of high concentration and a living organism somewhere else. The estimation of contaminant concentrations in such pathways, and therefore the hazards they create, demands careful consideration, particularly since marine food chains are often long (5 trophic levels) compared with terrestrial ones (3 trophic levels).

A concentration factor approach is adequate for living organisms which are more or less stationary and remain in one area where contaminant concentrations may be homogeneous. This assumes that the contaminants concentration in the organism is proportional to that in the water in the vicinity in which it lives, so that if the water concentration doubles, so (eventually) will the concentration in the organism. It does not matter whether the contaminant actually reaches the organism indirectly (e.g. via sediment or plant material) or not, so long as there is a one-to-one correspondence in concentrations over some appropriate averaging time. The exact mechanisms need not be studied provided field data for the actual concentration factor can be obtained. Such stationary biota may thus be treated as part of the marine environment, since the contaminant concentrations in them may be directly estimated using an appropriate concentration factor from the local water concentration.

This simple method does not apply to organisms which undertake major migrations across regions of high concentration gradient (e.g. in the vertical, perhaps). Their contaminant concentration cannot be simply related to the water concentration in the region where they are likely to be found (and eaten), since they may have been feeding elsewhere where the concentration is quite different. It could also be related to the average concentration experienced during the migratory movement, if they are absorbing contaminant directly from the water. More detailed consideration of the mechanisms involved is therefore required in this case.

Studies of concentration factors have shown that concentration of certain contaminants increases with increasing steps in the food chain although in many cases the opposite effect occurs. There have been many attempts to construct food chain models which show this effect – for example that of Williams (1972) for steady-state, non-linear food webs; three- and four-compartmental models (Conover and Francis, 1973); the use of gamma and log-normal distributions (Eberhardt and Gilbert, 1973); equilibrium models for diverse food chains (Thomann, 1981); and models specifically designed to estimate the transfer of radionuclides from the deep-sea disposal of waste to man (Doi et al., 1980).

1. CONCENTRATION BY ORGANISMS

At the other extreme from such detailed approaches are studies of
the transfer of radionuclides or contaminants which undergo degradation
or change from one link in the food chain. Simple expressions may be
used to represent these transfers (Pentreath, 1983).

Consider an organism of weight W feeding, at a fractional rate f of
its body weight per unit time, on food containing a contaminant at
concentration γ. Its ingestion rate I of contaminant (quantity per
unit time) is therefore

$$I = fW\gamma \tag{A.X.1}$$

and if it only retains in its body a fraction r of what it ingests, its
effective intake rate is

$$I = rfW\gamma \tag{A.X.2}$$

Suppose that it also eliminates, by excretion or other losses, a
fraction k of its body burden Q of contaminant per unit time. Intake
must balance elimination at steady state, so that equilibrium body burden
Q_{ss} is given by

$$kQ_{ss} = rfW\gamma \tag{A.X.3}$$

In the special case where a contaminant is accumulated without
elimination, k may be replaced by the reciprocal of the mean lifetime of
the organism. In this modified form Equation (A.X.3) would then simply
state that the ultimate body burden is the accumulation of all
contaminants taken in during the lifetime.

The whole-body concentration in the organism is Q_{ss}/W, so the
concentration of contaminant in the organism, relative to that in its
food (the upchain <u>transfer factor</u>) denoted by β, is

$$\beta = \frac{Q_{ss}/W}{\gamma} = \frac{rfW\gamma}{kW\gamma} = \frac{rf}{k} \tag{A.X.4}$$

The ratio $\beta = rf/k$ encapsulates all the essential factors which
determine upchain concentration of contaminants. It has been derived
here in its simplest form for a stable contaminant but various features
specific for different types of contaminant may be incorporated. For
instance, to include radioactive decay, the elimination rate is modified
by adding the radioactive decay constant, and therefore replacing k by
$k + \lambda$. The essential features remain unchanged by such elaborations,
which are discussed by Pentreath (1983). One particularly important
modification is to allow for direct input from water, which may be the
most significant pathway of uptake for certain substances, so that the
local concentration factor method discussed above can be applied.

Since r, the fraction retained, cannot exceed one, it is clear that
the transfer factor only exceeds one if the fractional feeding rate
exceeds the fractional elimination rate. This may of course occur and
there is therefore no a priori reason to suppose that transfer factors
are greater or less than one. Successive concentration of contaminants
as they move up food chains is <u>not</u> a universal rule. Many cases of

decreasing concentrations have been observed. In any case, an absolute upper limit to the whole-body transfer factor is given by the total retention (zero elimination) case, when β reduces to the number of times that an organism consumes its final body weight in its lifetime. This is not generally a very large number (between 3 and 30, perhaps), although for birds and mammals it is typically 100.

Contaminants are usually found concentrated in particular organs, so that the transfer factor for that organ (related to average food concentration) will be higher by a factor g/w, where g is the fraction of the body burden contained in the organ, and w is the fraction that that organ represents of whole-body weight. This does not affect the conclusion above, however, unless predators feed selectively on particular organs. Most marine predators in fact ingest whole bodies, so that the elevated organ concentrations are relevant only at the final stage, where consumption by man is often of particular organs or tissues. These are of course quite likely to be organs (e.g. muscle) in which concentrations are depressed relative to the whole-body level, rather than those in which they are elevated. Another situation arises with the high concentrations of a contaminant in eggs that may lead to exceptionally high levels in fry and larva - a form of inherited bioaccumulation. In general this will be diluted with growth.

2. MULTIPLE LINK FOOD CHAINS

The computation of concentration in a foodstuff may therefore proceed in several stages:

(1) Compute concentration in local (sedentary) food using an appropriate concentration factor and the water or sediment concentration.
(2) Compute concentrations up the food chain (possibly several links) using whole-body transfer coefficients.
(3) Compute concentration in final foodstuff using appropriate (maybe single-tissue) concentration elevation factor.

In many cases the intermediate step (2) is unnecessary. Where required, the calculations may either increase or decrease concentrations, and where they do increase them, the factor involved is unlikely to be more than a few orders of magnitude at most. In general the principal concentration will occur at steps (1) and (3), which are no different than those for single-link food chains. Generally speaking, therefore, there is no fundamental reason to suppose that multiply linked (long) food chains will lead to higher exposures than short ones. The reverse may very likely be true. The principal importance of long food chains may rather be that they provide a direct link with an otherwise inaccessible region of high concentration.

The calculation outlined above also assumes that the original food in a region of high concentration forms the entire diet of the animals feeding on it, that they in turn comprise the exclusive diet of the next animals in the chain, and so on. This is of course unrealistic. At each stage the animals are likely to range more or less widely, and the contaminated part of their diet is diluted with less highly contaminated food from elsewhere. There are two ways to allow for this. Firstly one could estimate the probability of eating contaminated food at each stage and multiply the transfer factor of Equation (A.X.4) by this probability (Pentreath, 1983). Alternatively, one may estimate the average concentration in the food eaten at each stage. In either case some knowledge of the feeding range of the animals is required.

These calculations apply only when intake of contaminant is entirely via food. In cases where intake via water ingested, or by direct absorption through the skin, across gills, etc., occurs, an appropriate concentration factor relative to the average water concentration experienced by an organism can be estimated, either from field observations or from similar calculations to those above. For fat-soluble compounds specifically a solution equilibrium between body fat and surrounding water may provide a mechanism for excretion as well as accumulation.

For contaminant transfer models in general it should also be noted that, although some allowance is sometimes made for degradation – such as isotopic decay – allowance is not usually made for transformation, such as inorganic to organic forms of mercury. Even fairly sophisticated models of methylmercury transfer through aquatic food chains (e.g. those by Fagerstrom and Asell, 1973; Fagerstrom, Asell and Jernelov, 1974) assume that no conversion of one form to another takes place. The relative merits of the different approaches made to modelling the upper end of the food chain for radionuclide transfer have been briefly discussed in a review by Pentreath (1977).

The probability of food chain transfers which link the deep-sea organisms to food consumed by humans is low, but since such exposure may occur it may be useful to calculate limiting cases. The most likely food chains leading to high exposure in practice are likely to arise in coastal waters, where by far the vast majority of marine food products are harvested. Considerable reliance would therefore have to be placed on model predictions of coastal water contaminant concentrations. In these cases concentration factor data are available and could be used. Unfortunately direct comparisons of different model predictions with actual data are not yet available. It can readily be envisaged, however, that for a highly soluble radionuclide such as ^{99}Tc, direct food chain transfer is unlikely to produce a higher concentration in a predator fish caught in surface coastal waters than the concentration attained, by direct adsorption, by coastal water algae because of the latter's ability to concentrate this nuclide (Pentreath et al., 1981). The relative importance of hypothetical critical pathways will clearly be governed by the combinations of type of food eaten by man (fish, shellfish, algae), the quantities of each eaten, and the affinities of each type of foodstuff for the contaminant. It is therefore worth recalling that all algae live in waters of $<$ 100 m depth (and algae frequently have the highest concentration factors for many elements), that the majority of shellfish (molluscs and crustaceans) liable to be consumed at a sustained rate will also be taken from coastal waters, and that neither of these two foodstuffs will be linked by a predator/prey food chain to the deep ocean. The presence of macrophyte (seagrass) debris in the deep sea (Wolff, 1976) indicates that some (downward) transport between continental shelf and abyssal regions does occur, but the mass involved is quantitatively insignificant.

The only food chain pathway of real concern, therefore, is that which results in the consumption of fish. Fish are eaten in large quantities but tend to have low concentration factors for most elements and other potential contaminants – with the notable exceptions of Cs and Hg. This is partly the result of the fact that a large fraction of their body weight consists of muscle. The general importance of relating organ concentrations to those of whole body, and to food eaten, has been discussed above (see also Pentreath, 1977) but it is a factor which should be taken into account. In view of the above, the inference is that food chains resulting in the exposure of critical groups, along critical pathways – are most likely to arise in coastal waters. The greatest biological need in 'modelling' this aspect of dumping is the procurement of better concentration factor data applicable to all marine species already being consumed from coastal water fisheries.

REFERENCES

ALLEN, C. M., 1982. Numerical simulation of contaminant dispersion in
estuary flows. Proc. R. Soc. London A 381, 179-194.

AMIN, B. S., S. KRISHNASWAMY and B. L. K. SOMAYAJULU, 1974.
$^{234}Th/^{238}U$ activity ratios in Pacific Ocean bottom waters. Earth
Planet. Sci. Lett. 21, 342-344.

ANDERSON, D. R., Editor, 1982. Seventh International OECD Nuclear Energy
Agency/Seabed Working Group Meeting, La Jolla, California, March 15-19,
1982. Sandia Report SAND 82-0460, 221 pp.

ANDERSON. R. F., 1981. The marine geochemistry of thorium and
protactinium. Ph. D. Thesis, Massachusetts Institute of Technology/Woods
Hole Oceanographic Institute Joint Program, Woods Hole, Mass., 287 pp.

ANGEL, M. V. and A. de C. BAKER, 1982. Vertical distribution of the
standing crop of plankton and micronekton at three stations in the
Northeast Atlantic. Biol. Oceanog. 2, 1-30.

ANGEL, M. V., 1983. Detrital organic fluxes through pelagic ecosystems.
In: Flows of Energy and Materials in Marine Ecosystems: Theory and
Practice, M. J. Fasham, Editor, Plenum Press (NATO Science Series).

BACON, M. P., D. W. SPENCER and P. G. BREWER, 1976. $^{210}Pb/^{226}Ra$
and $^{210}Po/^{210}Pb$ disequilibria in seawater and suspended particulate
matter. Earth Planet. Sci. Lett. 32, 277-296.

BACON, M. P. and R. F. ANDERSON, 1982. Distribution of thorium isotopes
between dissolved and particulate forms in the deep sea. J. Geophys.
Res. 87, 2045-2056.

BENNETT, A. F. and D. B. HAIDVOGEL, 1981. Low resolution numerical
simulation of damped two-dimensional turbulence. Ocean Modelling, 40,
5-10, available from Editor: Robert Hooke Institute, Department of
Atmospheric Physics, Parks Road, Oxford OX1 3PU, England, on behalf of
SCOR.

BHAT, S. G., S. KRISHNASWAMY, D. LAL and RAMA and W. S. MOORE, 1969.
$^{234}Th/^{238}U$ ratios in the ocean. Earth Planet. Sci. Lett. 5, 483-491.

BLACK, R. and D. B. BOUDRA, 1981. Initial testing of a numerical ocean
circulation model using a hybrid (quasi-isopycnic) vertical coordinate,
J. Phys. Oceanog. 11, 755-770.

BOWEN, V. T., V. E. NOSHKIN, H. D. LIVINGSTON and H. L. VOLCHOK, 1980.
Fallout radionuclides in the Pacific Ocean vertical and horizontal
distributions, largely from GEOSECS stations. Earth Planet. Sci. Lett.
49, 411-434.

BREWER, P. G., D. W. SPENCER, P. E. BISCAYE, A. HANLEY, P. L. SACHS,
C. L. SMITH, S. KADAR, and J. FREDERICS, 1976. The distribution of
particulate matter in the Atlantic Ocean. Earth Planet. Sci. Lett. 32,
393-402.

BREWER, P. G., Y. NOZAKI, D. W. SPENCER and A. P. FLEER, 1980. Sediment trap experiments in the deep North Atlantic: isotopic and elemental fluxes. J. Mar. Res. 38, 703-728.

BROECKER, W. S., A. KAUFMAN and R. M. TRIER, 1973. The residence time of thorium in surface sea water and its implications regarding the fate of reactive pollutants. Earth Planet. Sci. Lett. 20, 35-44.

BRULAND, K. W., 1980. Oceanographic distributions of cadmium, zinc, nickel and copper in the North Pacific. Earth Planet. Sci. Lett. 47, 176-198.

CARPENTER, R., J. T. BENNETT and M. L. PETERSON, 1981. ^{210}Pb activities in and fluxes to sediments of the Washington continental slope and shelf. Geochim. Cosmochim. Acta 45, 1155-1172.

CHUNG, Y. and H. CRAIG, 1980. ^{226}Ra in the Pacific Ocean. Earth Planet. Sci. Lett. 49, 267-292.

CLARKE, M. R., 1977. Beaks, nets and numbers. The biology of cephalopods. In: The Biology of Cephalopods, M. Nixon and J.B. Messenger, Editors, (Symp. Zool. Soc. Lond. 38,) Academic, London pp. 89-126.

COCHRAN, J. K., 1982. The oceanic chemistry of the U- and Th-series nuclides. In: Uranium Series Disequilibrium - Applications to Environmental Problems, M. Ivanovich and R. S. Harmon, Editors, Oxford Univ. Press, London, pp. 384-430.

CONOVER, R. J., and V. FRANCIS, 1973. The use of radioactive isotopes to measure the transfer of materials in aquatic food chains. Mar. Biol. 18, 272-283.

Convention on the Prevention of Marine Pollution by Dumping of Wastes and Other Matter, London, 1972, reprinted in Inter-Governmental Conference on the Convention on the Dumping of Wastes at Sea, 1982 edition, International Maritime Organization, London, pp. 7-16.

CRAIG, H., S. KRISHNASWAMY and B. L. K. SOMAYAJULU, 1973. ^{210}Pb-^{226}Ra: radioactive disequilibrium in the deep sea. Earth Planet. Sci. Lett. 17, 295-305.

CRAIG, H., 1974. A scavenging model for trace elements in the deep sea. Earth Planet. Sci. Lett. 23, 149-159.

CRANSTON, R. E. and J. W. MURRAY, 1978. The determination of chromium species in natural waters. Anal. Chim. Acta 99, 275-282.

CSANADY, G. T., 1973. Turbulent Diffusion in the Environment. D. Reidel Pub. Co. Dordrecht, Netherlands, 248 pp.

CUSHING, D. H., 1971. Upwelling and the production of fish, Adv. Mar. Biol. 9, 255-334.

DeMASTER, D. J., 1979. The marine budgets of silica and ^{32}Si. Ph. D. Thesis, Yale Univ., New Haven, Conn., 324 pp.

150

DEMING, J. W., 1981. Ecology of barophilic deep-sea bacteria. Ph.D. Thesis, University of Maryland, College Park, Md., 160 pp.

DICKINSON, R. R., 1983. Global summaries and intercomparisons, flow statistics from long-term current meter moorings. Chapter 15 in Eddies in Marine Science, A.R. Robinson, Editor, Springer-Verlag, Berlin, Heidelberg, New York.

DOI, T., T. KIDACHI, K. HONJO, Y. MATSUSHITA, T. NEMOTO, M. SHIMIZU, H. SUDO and H. TSURUGA, 1980. A preliminary assessment of biological transport of radionuclides dumped at deep sea bottom. In: Marine Radioecology, Proc. 3rd NEA Seminar on Marine Radioecology, Tokyo, Japan, Oct. 1-5, 1979, NEA/OECD, Paris, 95-110.

EBERHARDT, L.L., and R.O. GILBERT, 1973. Gamma and lognormal distributions as models in studying food-chain kinetics. Battelle Pacific Northwest Labs. Report BNWL-1747, 100 pp.

FAGERSTROM, T., and B. ASELL, 1973. Methyl mercury accumulation in an aquatic food chain: A model and some implications for research planning. Ambio 2, 164-171.

FAGERSTROM, T., B. ASELL and A. JERNELOV, 1974. Model for accumulation of methyl mercury in northern pike Esox lucius. Oikos 25, 14-20.

FAO, 1980. 1979 Yearbook of Fisheries Statistics, V. 48, Food and Agriculture Organization of the United Nations, Rome. 384 pp.

FIADEIRO, M. E., and H. CRAIG, 1978. Three dimensional modeling of tracers in the deep Pacific Ocean: I. Salinity and oxygen. J. Mar. Res. 36, 323-355.

FOWLER, S. W., 1982. Biological transfer and transport processes. In: Pollutant Transfer and Transport in the Sea, G. Kullenberg, Editor. Vol. II, C.R.C. Press Inc., Boca Raton, Florida, p. 1-65.

GARRETT, C. J. R., 1981. Streakiness. Ocean Modelling 41, 4-6, available from Robert Hooke Institute, Oxford, England.

GARRETT, C. J. R., 1983. On the initial streakiness of a dispersing tracer in two- and three-dimensional turbulance. Dynamics of Atmospheres and Oceans 7, 265-277.

GARRETT, C. J. R. and J. R. SHEPHERD, 1986. A simple model for pollutant dispersal from a source in a scavenging ocean (in preparation).

GESAMP (IMCO/FAO/UNESCO/WMO/WHO/IAEA/UN/UNEP Joint Group of Experts on the Scientific Aspects of Marine Pollution) 1982. The Review of the Health of the Oceans. GESAMP Reports and Studies No.15, Unesco, Paris, 108 pp.

HAEDRICH, R. L. and G. T. ROWE, 1977. Megafaunal biomass in the deep sea. Nature (London) 269, 141-142.

HAEDRICH, R. L., G. T. ROWE and P.T. POLLONI, 1980. Megabenthic fauna in the deep sea south of New England, USA. Mar. Biol. 57, 165-179.

HESSLER, R., 1981. Oasis under the sea - where sulphur is the staff of life. New Scientist 92, 741-742, 744-747.

HINGA, K. R., J. McN. SIEBURTH and G. R. HEATH, 1979. The supply and use of organic material at the deep-sea floor. J. Mar. Res. 37, 557-579.

HOLLAND, W. R. and A. D. HIRSCHMAN, 1972. A numerical calculation of the circulation of the North Atlantic Ocean. J. Phys. Oceanog. 2, 336-354.

HOLLOWAY, G., 1982. Comments on streakiness. Ocean Modelling 43, 5-6, available from Robert Hooke Institute, Oxford, England.

HONJO, S., S. J. MANGANINI and J. J. COLE, 1982. Sedimentation of biogenic matter in the deep ocean. Deep-Sea Res. 29A, 609-625

INTERNATIONAL ATOMIC ENERGY AGENCY, 1976. Effects of Ionizing Radiation on Aquatic Organisms and Ecosystems. Technical Reports Series No. 172, IAEA, Vienna, 131 pp.

INTERNATIONAL ATOMIC ENERGY AGENCY, 1978. The oceanographic basis of the IAEA revised definition and recommendations concerning high-level radioactive waste unsuitable for dumping at sea. TECDOC IAEA-210, International Atomic Energy Agency, Vienna, 59 pp.

INTERNATIONAL ATOMIC ENERGY AGENCY, 1979. Methodology for Assessing Impacts of Radioactivity on Aquatic Ecosystems. Technical Reports Series No. 190, IAEA, Vienna, 416 pp.

ISAACS, J. D., 1972. Unstructured marine food webs and "pollutant analogues". Fish. Bull. 70, 1053-1059.

ISAACS, J. D., 1973. Potential trophic biomasses and trace-substance concentrations in unstructured marine food webs. Mar. Biol. 22, 97-104.

ISAACS, J. D. and R. SCHWARTZLOSE, 1975. Active animals of the deep-sea floor. Sci. Am. 233 (4), 85-91.

KADKO, D. C., 1981. A detailed study of uranium series nuclides for several sedimentary regimes of the Pacific. Ph. D. Thesis, Columbia Univ., New York, 330 pp.

KEFFER, T. and D. B. HAIDVOGEL, 1982. Numerical simulations of tracer streakiness. Ocean Modelling 45, 1-4, available from Robert Hooke Institute, Oxford, England.

KLINKHAMMER, G. P., and M. L. BENDER, 1980. The distribution of manganese in the Pacific Ocean. Earth Planet. Sci. Lett. 46, 361-384.

KUO, H. H. and G. VERONIS, 1973. The use of oxygen as a test for an abyssal circulation model. Deep-Sea Res. 20, 871-888.

KUPFERMAN, S., and D. E. MOORE, 1981. Physical oceanographic characteristics influencing the dispersion of dissolved tracers released at the sea floor in selected deep ocean study areas. Sandia Report SAND 80-2573, 28 pp.

LAL, D., 1980. Comments on some aspects of particulate transport in the oceans. Earth Planet. Sci. Lett. 49, 520-527.

LI, Y. H., H. W. FEELY and P. S. SANTSCHI, 1979. ^{228}Th-^{228}Ra radioactive disequilibrium in the New York Bight and its implications for coastal pollution. Earth Planet. Sci. Lett. 42, 13-26.

LI, Y. H., J. R. TOGGWEILER and H. W. FEELY, 1980. ^{228}Ra and ^{228}Th concentrations in GEOSECS Atlantic surface waters. Deep-Sea Res. 27A, 545-555.

LI, Y. H., P. H. SANTSCHI, A. KAUFMAN, L. K. BENNINGER and H. W. FEELY, 1981. Natural radionuclides in waters of the New York Bight. Earth Planet. Sci. Lett. 55, 217-228.

LIVINGSTON, H. D., and V. T. BOWEN, 1979. Pu and ^{137}Cs in coastal sediments. Earth Planet. Sci. Lett. 43, 29.

LONGHURST, A.R., 1976. Vertical migration. In: The Ecology of the Seas, D. H. Cushing, J. J. Walsh, Editors, W. B. Saunders Co., Philadelphia, Toronto, 116-137.

LUPTON, J. E., and H. CRAIG, 1981. A major He^3 source at 15°S on the east-Pacific Rise, Science 214, pp. 13-14.

MARSHALL, N. B. and N. R. MERRETT, 1977. The existence of a benthopelagic fauna in the deep-sea. In: A Voyage of Discovery, M. V. Angel, Editor, Pergamon Press, Oxford, 483-497.

MARSHALL, N.B., 1979. Developments in Deep-Sea Biology, Blandford Press, Poole, 566 pp.

MATSUMOTO, E., 1975. ^{234}Th-^{238}U radioactive disequilibrium in the surface layer of the ocean. Geochim. Cosmochim. Acta 39, 205-212.

MEASURES, C. I., R. E. McDUFF and J. M. EDMOND, 1980. Selenium redox chemistry at GEOSECS I re-occupation. Earth Planet. Sci. Lett. 49, 102-108.

MOORE, W. S., K. W. BRULAND and J. MICHEL, 1981. Fluxes of uranium and thorium series isotopes in the Santa Barbara Basin. Earth Planet. Sci. Lett. 53, 391-399.

McDOWELL, S. E. and H. T. ROSSBY, 1978. Mediterranean water: an intense mesoscale eddy off the Bahamas. Science 202, 1085-1087.

McWILLIAMS, et al., 1983. Western North Atlantic - Local Dynamics of Eddies, Chapter 6. In: Eddies in Marine Science, A.R. Robinson, Editor, Springer-Verlag, Berlin, Heidelberg, New York.

NEA Seabed Working Group, 1982. Report of the Systems Analysis Task Group. Sandia Report SAND 82-0460.

NOSHKIN, V. E. and V. T. BOWEN, 1973. Concentrations and distributions of long-lived fallout radionuclides in open ocean sediments. In: Radioactive Contamination of the Marine Environment (Proc. Symp. Seattle, 1972), IAEA, Vienna, 671.

NOZAKI, Y., J. THOMSON and K. K. TUREKIAN, 1976. The distribution of ^{210}Pb and ^{210}Po in the surface waters of the Pacific Ocean. Earth Planet. Sci. Lett. 32, 304-312.

NOZAKI, Y. and S. TSUNOGAI, 1976. ^{226}Ra, ^{210}Pb and ^{210}Po disequilibria in the western North Pacific. Earth Planet. Sci. Lett. 32, 313-321.

NOZAKI, Y., Y. HORIBE and H. TSUBOTA, 1981. The water column distributions of thorium isotopes in the western North Pacific. Earth Planet. Sci. Lett. 54, 203-216.

OSTERBERG, C., A. G. CAREY and H. CURL Jr., 1963. Acceleration of sinking rates of radionuclides in the oceans. Nature (London) 200, 1276-1277.

PASQUILL, F., 1974. Atmospheric diffusion: the dispersion of windborne material from industrial and other sources (2nd edition). Ellis Horwood Ltd., Chichester, England, 429 pp.

PENTREATH, R. J., 1977. Radionuclides in marine fish. Oceanogr. Mar. Biol. Ann. Rev. 15, 365-460.

PENTREATH, R.J., 1980. Nuclear Power, Man and the Environment. Taylor and Francis, London.

PENTREATH, R. J., D. F. JEFFERIES, M. B. LOVETT and D. M. NELSON, 1980. The behaviour of transuranic and other long-lived radionuclides in the Irish Sea and its relevance to the deep sea disposal of radioactive wastes. In: Marine Radioecology, Proc. 3rd NEA Seminar on Marine Radioecology, Tokyo, Japan, Oct. 1-5, 1979. NEA/OECD, Paris, 203-221.

PENTREATH, R. J., 1981. The biological availability to marine organisms of transuranium and other long-lived nuclides. In: Impacts of Radionuclide Releases into the Marine Environment (Proc. Symp. Vienna, 1981), IAEA/NEA(OECD), Vienna, 241-272.

PENTREATH, R. J., 1983. Biological studies In: Interim Oceanographic Description of the NEA Dumpsite for the Disposal of Low-level Radioactive Waste. P.A. Gurbutt and R.R. Dickson, Editors, NEA/OECD, Paris.

REID, J. L. and R. J. LYNN, 1971. On the influence of Norwegian – Greenland and Weddell seas upon the bottom waters of the Indian and Pacific Oceans. Deep-Sea Res. 18, 1063-1088.

RICHARDSON, P. L., 1982. Western North Atlantic: A vertical section of eddy kinetic energy along 55° W. ICES Document C.M. 1982/C:21, 12 pp.

RICKER, W. E., 1969. Food from the sea. In: Resources and Man: a study and recommendations by the Committee of Resources and Man, NAS-NRC. Freeman & Co., Chicago, 87-108.

RILEY, G. A., 1970. Particulate organic matter in sea water. Adv. Mar. Biol. 8, 1-118.

ROBINSON, A. R., D. E. HARRISON, Y. MINTZ and A. J. SEMTNER, 1977. Eddies and the general circulation of an idealized ocean gyre: A wind and thermally driven primitive equation numerical experiment. J. Phys. Oceanog. 7, 182-207.

ROBINSON, A. R., D. E. HARRISON and D. B. HAIDVOGEL, 1979. Mesoscale eddies and general ocean circulation models. Dyn. of Atmospheres and Oceans 3, 143-180.

ROBINSON, A. R. and M. M. MULLIN, 1981. A model for physical-biological transfer. In: Biological and Related Chemical Research Concerning Subseabed Disposal of High-level Nuclear Waste: report of a workshop at Jackson Hole, Wyoming, January 12-16, 1981, M. M. Mullin and L. S. Gomez, Editors, Sandia Report SAND 81-0012, 29-32.

ROWE, G. T., P. T. POLLONI and S. G. HORNER, 1974. Benthic biomass estimates from the northwestern Atlantic Ocean and the northern Gulf of Mexico, Deep-Sea Res. 21, 641-650.

ROWE, G. T. and W. D. GARDNER, 1979. Sedimentation rates in the slope water of the northwest Atlantic Ocean measured directly with sediment traps. J. Mar. Res. 37, 581-600.

ROWE, G. T., 1981. The deep-sea ecosystem. In: Analysis of Marine Ecosystems, A. R. Longhurst, Editor, Academic Press, London, 235-267.

ROWE, G. T., 1983. Biomass production in the deep-sea macrobenthos. In: The Sea, Vol. 8, Wiley-Interscience, New York.

SANTSCHI, P. H., Y. H. LI, J. J. BELL, R. M. TRIER and K. KAWTALUK, 1980. Pu in coastal marine environments. Earth Planet. Sci. Lett. 51, 248-265.

SARMIENTO, J. L., H. W. FEELY, W. S. MOORE, A. F. BAINBRIDGE and W. S. BROECKER, 1976. The relationship between vertical eddy diffusion and buoyancy gradient in the deep-sea. Earth Planet. Sci. Lett. 32, 357-370.

SARMIENTO, J. L. and K. BRYAN, 1982. An ocean transport model for the North Atlantic, J. Geophys. Res. 87, 394-408.

SCHAULE, B. and C. C. PATTERSON, 1980. The occurrence of lead in the northeast Pacific and the effects of anthropogenic inputs. In: Lead in the Marine Environment, M. Branica and Z. Konrad, Editors, Proceedings of the International Experts Discussion on Lead Occurrence, Fate and Pollution in the Marine Environment, Rovini, Yugoslavia, 18-22 October, 1977. Pergamon Press, Oxford.

SCHMITZ, W. J., Jr., and W. R. HOLLAND, 1982. A preliminary comparison of selected numerical eddy-resolving general circulation experiments with observations. J. Mar. Res. 40, 75-117.

SCLATER F. R., E. BOYLE and J. M. EDMOND, 1976. On the marine geochemistry of nickel. Earth Planet. Sci. Lett. 31, 119-128.

SEMTNER, A. J., 1974. An oceanic general circulation model with bottom topography. Univ. California, Los Angeles, Dept. Meteorology, Numerical Simulation of Weather and Climate, Tech. Report No. 9, 99 pp.

SEMTNER, A. J., and Y. MINTZ, 1977. Numerical simulation of the Gulf Stream and mid-ocean eddies. J. Phys. Oceanog. 7, 208-230.

SHEPHERD, J. G., 1976. A simple model for the dispersion of radioactive wastes dumped on the deep sea bed. Fisheries Research Technical Report 29. Ministry of Agriculture, Fisheries and Food, Lowestoft, UK, 19 pp.

SHEPHERD, J. G., 1980. A two-dimensional model of the meridional transport of tracers in the deep ocean. Ocean Modelling 31, 8-12, available from Robert Hooke Institute, Oxford, England.

SHEPHERD, J. G., 1983. A two-dimensional meridional model of the dispersion of contaminants in the deep ocean: preliminary results, MAFF Fisheries Lab. Internal Report No. 8.

SHOLKOVITZ, E. R., 1983. The geochemistry of plutonium in fresh and marine water environments. Earth Sci. Rev. (in press).

SMITH, K. L., Jr., 1978. Metabolism of the abyssopelagic rattail Coryphaenoides armatus measured in situ. Nature (London) 274, 362-364.

SPENCER, D. W., M. P. BACON and P. G. BREWER, 1981. Models of the distribution of ^{210}Pb in a section across the North Equatorial Atlantic Ocean. J. Mar. Res. 39, 119-138.

THIEL, H., 1975. The size structure of the deep-sea benthos, Int. Revue Ges. Hydrobiol. 60, 575-606.

THOMANN, R. V., 1981. Equilibrium model of fate of microcontaminants in diverse aquatic food chains. Can. J. Fish. Aquat. Sci. 38, 280-296

THOMSON, J. and K. K. TUREKIAN, 1976. ^{210}Po and ^{210}Pb distributions in ocean water profiles from the eastern South Pacific. Earth Planet. Sci. Lett. 32, 297-303.

TUREKIAN, K. K., J. K. COCHRAN, L. K. BENNINGER and R. C. ALLER, 1980. The sources and sinks of nuclides in Long Island Sound. Adv. In Geophysics 22, 129-164.

VERONIS, G., 1975. The role of models in tracer studies. In: Numerical Models of Ocean Circulation: Proceedings of a symposium. US Natl. Acad. Sci., Washington, DC, 133-146.

VOORHIS, A. D. and D. C. WEBB, 1970. Large vertical currents observed in a winter sinking region of the Northwestern Mediterranean. Cahiers Océanographiques 22, 571-580.

WATSON, S. W., T. J. NOVITSKY, H. L. QUINBY and F. W. VALOIS, 1977. Determination of bacterial number and biomass in the marine environment. Appl. Environ. Microbiol. 33, 940-946.

WEBB, G. A. M. and P. D. GRIMWOOD, 1976. A revised oceanographic model to calculate the limiting capacity of the ocean to accept radioactive waste. UK National Radiological Protection Board Report No. NRPB-58, 19 pp.

WILLIAMS, R. B., 1972. Steady-state equilibriums in simple nonlinear food webs. In: Systems Analysis and Simulation in Ecology, B.C. Patten, Editor, Academic Press, New York, Vol.II, 213-240

WILLIAMS, P. M. and A. F. CARLUCCI, 1976. Bacterial utilization of organic matter in the deep sea. Nature (London) 262, 810-811.

WISHNER, K. F., 1980. Biomass of the deep-sea benthopelagic plankton. Deep-Sea Res. 27A, 203-216.

WOLFF, T., 1976. Utilization of seagrass in the deep sea. Aquat. Bot. 2, 161-174.

WUNSCH, C., 1981. New methods for diagnostic and box models. Ocean Modelling 40, 1-3, available from Robert Hooke Institute, Oxford, England.

WUNSCH, C., and J. F. MINSTER, 1982. Methods for box models and ocean circulation tracers: mathematical programing and nonlinear inverse theory. J. Geophys. Res. 87, 5647-5662.

GLOSSARY

This list provides guidance on a selection of technical terms and on some everyday words which are used in a specialized sense in the report

ADSORPTION/DESORPTION
: The process of attachment onto and release from particle surfaces

ANOXIC
: Devoid of free oxygen

ANTHROPOGENIC
: Man-made

AUTHIGENIC (of a mineral)
: Formed by sedimentary processes as a crystallographic unit at the place of its occurrence

AUTOCHEMOTROPHIC
: Pertaining to the synthesis of organic compounds from inorganic substance without need of sunlight

BENTHIC
: Living on or in the seabed

BIOGENIC
: Originating from biological processes

BIOMASS
: Total weight of organisms inhabiting a defined area

BIOTA
: The living things of a region

BIOTURBATION
: Disturbance of the surface sediments of the seabed caused by animal activity

BOUNDARY SCAVENGING
: Localization of the scavenging process at the oceanic margins such as the sea floor, pronounced bathymetric features like oceanic ridges or, particularly, the nearshore shelf-slope regime

COPROPHAGOUS
: Feeding on excrement

CORIOLIS EFFECT
: The deflection relative to the earth's surface of any object moving above the earth, caused by the Coriolis force; an object moving horizontally is deflected to the right in the northern hemisphere, to the left in the southern.

DEMERSAL
: Living on or near the seabed

DETRITIVOROUS
: Feeding on organic debris

DIAGENESIS
: Chemical and physical changes occurring in sediments during and after their deposition but before consolidation

159

DIAPYCNAL (PROCESS)	Pertaining to the transfer across isopycnal surfaces
DOUBLE DIFFUSION	A diffusive process that can arise from the different molecular transfer rates of heat and salt
ECOSYSTEM	A community of interdependent organisms together with their environment with which they interact
EKMAN LAYER	The layer that often exists at oceanic boundaries when frictional and Coriolis effects are dominant
EUPHOTIC ZONE	The upper level of the sea down to the limits of effective light penetration for photosynthesis
EUTROPHIC	Pertaining to a plentiful food supply
FAECAL WASTE, PELLETS, ETC.	Excrement
FOOD CHAIN	A number of organisms forming a feeding series through which energy is passed
GONAD	A primary sex gland; an ovary or a testis
HADAL	Pertaining to depths greater than 6000 m
ISOPYCNAL	Line of constant density
KINETICS	The rates of processes, specifically of uptake and release of chemical elements on particles
LARVA	The pre-adult form in which some animals hatch from the egg
MACROPHYTIC ALGAE	The larger seaweeds
METABOLISM	Chemical processes in general which occur within an organism, or part of one
OLIGOTROPHIC	Pertaining to a low food supply
ONTOGENETIC	Pertaining to the whole course of development during the life history of an individual
ORGANISM	Anything capable of carrying on life processes
PARAMETRIZATION	The simplification of the equations representing a system by mathematically describing some processes (usually of small scale) in reduced detail (often in terms of large-scale quantities)

PARTICLE-REACTIVE	Describes any chemical element or species which interacts with particles in the oceans
PELAGIC	Living above the seabed
PELAGIC OOZE	A deposition on the seabed of hard parts of dead pelagic organisms
PHOTOSYNTHESIS	The combining of water and carbon dioxide by green plants to form organic compounds using energy absorbed from sunlight
PHYTOPLANKTON	Drifting aquatic plants
PLANKTON	Aquatic organisms which drift with the currents
PYCNOCLINE	A density gradient - see thermocline
RADIOACTIVE EQUILIBRIUM	The situation existing when the local activity of a radionuclide equals that of its parent. "Disequilibrium" refers to perturbation in this state produced by geochemical processes
RESIDENCE TIME	A time characteristic of the length of time spent by a substance in an oceanic system. The definition of a single residence time useful for describing the fate of a contaminant in the ocean is often inappropriate.
SCAVENGING	The removal of chemical elements from the ocean by their incorporation into or attachment onto surfaces of particles
SESSILE	Fixed in one position
SINK	Area(s) in the oceans where particle-reactive chemical elements are deposited after scavenging
SPECIATION (of a chemical)	Refers to the various forms in which a chemical can exist
TERRIGENOUS	Originating from the land
TEST	A shell or hardened outer covering
THERMOCLINE	A temperature gradient, as in a layer of sea water, in which the temperature decrease with depth is greater than that of the overlying and/or underlying water
THERMOHALINE CIRCULATION	That part of the ocean circulation driven by changes in surface density or sinking water masses

TRACER	An oceanic property the conservation of which is well enough understood so that its distribution may be used to deduce oceanic transport processes
TROPHIC	Pertaining to nutrition
UPWELLING	The process by which water rises from a deeper to a shallower depth
URANIUM AND THORIUM DECAY SERIES	Chains of radioactive transformations headed by the naturally occurring isotopes ^{238}U, ^{235}U and ^{232}Th
WATER MASS	An oceanic water type, usually with a well-defined characteristic temperature and salinity

LIST OF BASIC SYMBOLS USED

C	=	concentration in water (/volume)
C_p	=	concentration in particles (/volume)
C_s	=	concentration in sediment (/volume)
f	=	volume of particles/unit volume of water column
f'	=	volume of sedimentary material/unit volume of sedimentary layer
h	=	depth of bioturbated or well-mixed sedimentary layer
K_b	=	effective vertical diffusivity in bioturbated layer
K_d	=	concentration in sediments per unit volume/concentration in water per unit volume
K_H	=	horizontal eddy diffusivity in water column
K_{pw}	=	pore-water vertical diffusivity
K_V	=	vertical eddy diffusivity in water column
k_1, k_2	=	rate constants (bulk) - water phase
k_1^s, k_2^s	=	rate constants (bulk) - sediment phase
L, H	=	width (length), depth of ocean
Q	=	release rate into the ocean
U	=	speed of mean flow
$u, v, w†$	=	velocity of water
V	=	volume of ocean
V_d	=	deposition velocity
V_i	=	flux of contaminant per unit water concentration by sinking particles
w_p	=	vertical speed of particles (+ = up, − = down)
w_s	=	net sediment accumulation rate
$x, y, z†$	=	co-ordinates (origin on bottom)
γ	=	concentration in biomass per unit volume/concentration in environment per unit volume
δ	=	fraction of particulate matter reaching bottom which does not dissolve in sedimentary layer
λ	=	radioactive decay rate

163

LIST OF WORKING GROUP MEMBERS, SECRETARIAT AND MEETINGS

Working Group Members

G.T. Needler   (Chairman)
Atlantic Oceanograhic Laboratory
Bedford Institute of Oceanography
P.O. Box 1006
Dartmouth, NS, B2Y 4A2
Canada

J.K. Cochran
Woods Hole Oceanographic Institution
Woods Hole, MA 02543
USA

J. Edmond
Massachusetts Institute of Technology
Dept. of Earth & Planetary Sciences
PM 54-1314
Cambridge, MA 02139
USA

C.J.R. Garrett
Department of Oceanography
Dalhousie University
Halifax, NS, B2H 4JI
Canada

G. Kullenberg
University of Copenhagen
Institute of Physical Oceanography
Haraldsgade 6
DK-2200 Copenhagen
Denmark

N.R. Merrett
Institute of Oceanographic Sciences
Brook Road
Wormley
Godalming, Surrey GU8 5UB
England

Y. Nozaki
Ocean Research Institute
University of Tokyo
1-15-1 Minamidai
Nakano-ku
Tokyo 164
Japan

G.T. Rowe
Ocean Science Division
Bldg. 318
Brookhaven National Laboratory
Upton, Long Island, NY 11973
USA

J.G. Shepherd
Fisheries Laboratory
Pakefield Road
Lowestoft, Suffolk NR33 OHT
England

S.A. Thorpe
Institute of Oceanographic Sciences
Brook Road
Wormley
Godalming, Surrey GU8 5UB
England

Secretariat

W.O. Forster   (Meeting 1)
J. Molinari   (Meetings 1-3)
A.A. Hagen   (Meetings 3-5)
International Atomic Energy Agency
P.O. Box 100,
Wagramerstrasse 5
A-1140 Vienna, Austria

M.K. Nauke   (Meetings 1-2)
International Maritime Organization
4 Albert Embankment
London SE1 7SR
England

Meeting Dates

(1) International Atomic Energy Agency, Vienna, 8-12 December 1980

(2) Woods Hole Oceanographic Institution, USA, 1-5 June 1981

(3) Principality of Monaco, 23-27 November 1981

(4) Bedford Institute of Oceanography, Canada, 3-7 May 1982

(5) International Atomic Energy Agency, Vienna, 22-26 November 1982

HOW TO ORDER IAEA PUBLICATIONS

 An exclusive sales agent for IAEA publications, to whom all orders and inquiries should be addressed, has been appointed in the following country:

UNITED STATES OF AMERICA   Bernan Associates — UNIPUB, 10033-F King Highway, Lanham, MD 20706

 In the following countries IAEA publications may be purchased from the sales agents or booksellers listed or through your major local booksellers. Payment can be made in local currency or with UNESCO coupons.

ARGENTINA	Comisión Nacional de Energía Atómica, Avenida del Libertador 8250, RA-1429 Buenos Aires
AUSTRALIA	Hunter Publications, 58 A Gipps Street, Collingwood, Victoria 3066
BELGIUM	Service Courrier UNESCO, 202, Avenue du Roi, B-1060 Brussels
CHILE	Comisión Chilena de Energía Nuclear, Venta de Publicaciones, Amunategui 95, Casilla 188-D, Santiago
CHINA	IAEA Publications in Chinese: China Nuclear Energy Industry Corporation, Translation Section, P.O. Box 2103, Beijing IAEA Publications other than in Chinese: China National Publications Import & Export Corporation, Deutsche Abteilung, P.O. Box 88, Beijing
CZECHOSLOVAKIA	S.N.T.L., Mikulandska 4, CS-116 86 Prague 1 Alfa, Publishers, Hurbanovo námestie 3, CS-815 89 Bratislava
FRANCE	Office International de Documentation et Librairie, 48, rue Gay-Lussac, F-75240 Paris Cedex 05
HUNGARY	Kultura, Hungarian Foreign Trading Company, P.O. Box 149, H-1389 Budapest 62
INDIA	Oxford Book and Stationery Co., 17, Park Street, Calcutta-700 016 Oxford Book and Stationery Co., Scindia House, New Delhi-110 001
ISRAEL	Heiliger and Co., Ltd, Scientific and Medical Books, 3, Nathan Strauss Street, Jerusalem 94227
ITALY	Libreria Scientifica, Dott. Lucio de Biasio "aeiou", Via Meravigli 16, I-20123 Milan
JAPAN	Maruzen Company, Ltd, P.O. Box 5050, 100-31 Tokyo International
NETHERLANDS	Martinus Nijhoff B.V., Booksellers, Lange Voorhout 9-11, P.O. Box 269, NL-2501 The Hague
PAKISTAN	Mirza Book Agency, 65, Shahrah Quaid-e-Azam, P.O. Box 729, Lahore 3
POLAND	Ars Polona-Ruch, Centrala Handlu Zagranicznego, Krakowskie Przedmiescie 7, PL-00-068 Warsaw
ROMANIA	Ilexim, P O. Box 136-137, Bucharest
SOUTH AFRICA	Van Schaik Bookstore (Pty) Ltd, P.O. Box 724, Pretoria 0001
SPAIN	Díaz de Santos, Lagasca 95, E-28006 Madrid Díaz de Santos, Balmes 417, E-08022 Barcelona
SWEDEN	AB Fritzes Kungl. Hovbokhandel, Fredsgatan 2, P.O. Box 16356, S-103 27 Stockholm
UNITED KINGDOM	Her Majesty's Stationery Office, Publications Centre, Agency Section, 51 Nine Elms Lane, London SW8 5DR
USSR	Mezhdunarodnaya Kniga, Smolenskaya-Sennaya 32-34, Moscow G-200
YUGOSLAVIA	Jugoslovenska Knjiga, Terazije 27, P.O. Box 36, YU-11001 Belgrade

---

 Orders from countries where sales agents have not yet been appointed and requests for information should be addressed directly to:

 Division of Publications
International Atomic Energy Agency
Wagramerstrasse 5, P.O. Box 100, A-1400 Vienna, Austria

86-02458